Ecosystem Modelling with EwE

Ecosystem Modelling with EwE

Villy Christensen and Carl J. Walters

THE UNIVERSITY OF BRITISH COLUMBIA
VANCOUVER, BC

Contents

Part V. Fitting Models to Data

Introduction

This textbook is really overdue. It's been planned for years, but as an undertaking without a deadline and a project without deliverables, it could be postponed till later. That's still the case, but now *is* the time, deliverables or no deliverables, time to just drop everything and write.

With the textbook, we provide a basis for courses in ecosystem modelling based on the Ecopath with Ecosim (EwE) modelling framework and software. EwE is a versatile approach with a low but very long (and gradually steeping) learning curve.

We have been involved in the development of EwE for more than thirty years, and during that time the simple Ecopath mass-balance approach it started with has through the addition of time and spatial-dynamic models, among others, been expanded to become a dynamic toolbox that can be used to address anything from simple fundamental research questions to very complex management and policy questions related to ecosystem-based management.

It's been a guiding principle throughout the development of EwE to provide an easy-to-access approach that does not require extensive mathematical or programming capabilities to get started. *Model first, ask later* is the philosophy. We think by just getting started with modelling, you gain insight and can start asking fundamental questions about how components of ecosystems interplay. Diving deeper is important, but we have sought to make the initial getting-started as simple and accessible as possible.

The textbook reflects our dedication to accessibility. We have chosen to develop an "open textbook", i.e. a textbook that is freely available on the web, that can be downloaded as a pdf, and also obtained as a printed textbook from booksellers in many countries. We are using the open book creation platform, PressBooks as implemented by BCcampus, which has several advantages. It is not only freely available, but also that it is straightforward to update the materials as new developments occur. Compared to printed, traditional textbooks, this is indeed one of the major advantages. We can also include media and interactive materials to liven up things a bit (in the online and pdf versions).

It's also important to us that the textbook is distributed with a Creative Commons (CC-BY-NC-SA) license. This makes it straightforward to adapt copies of the textbook, e.g., for you to use in your own courses. We indeed invite you to do so.

We have organized the textbook to focus on background, modelling aspects and to include tutorials with hands-on that illustrate many of the aspects (the "what" and "why"). The textbook is, however, not intended to be an in-depth description of everything in EwE. To supplement the textbook, with the "how", we refer to an accompanying publication, the EwE User Guide, which has been updated, overhauled, and also made accessible through BCcampus.

We have heard on many occasions that there is a strong, unfilled demand for training courses in ecosystem modelling with EwE. There are so many projects in progress around the world that includes EwE modelling, but there are very few instructors. While we have developed and shared course materials for years, there has never been a concerted action as represented by this textbook and the accompanying User Guide.

We hope both will be useful.

Foreword

Jeffrey Polovina

French Frigate Shoals (FFS) is a coral reef atoll located in the remote region of the Hawaiian Islands Archipelago. In the late 1970s it was the focus of marine scientists studying all components of the marine ecosystem from phytoplankton to seabirds and tiger sharks. In 1978 I was hired to model the energy flow for this coral reef ecosystem based on the findings from this research. While this was the age of single species models, I was aware of two efforts in temperate regions developing ecosystem models for well-studied systems. Unfortunately for us these models were very parameter-intensive, far beyond what we could estimate at FFS. Thus, I developed an approach termed as Ecopath to estimate the biomasses and the energy flow for the species groups that comprised the atoll ecosystem in our data-limited situation. The Ecopath model and its application to FFS was published in *Coral Reefs* in 1984.

While I was developing this model I received enthusiastic encouragement from Dr. Daniel Pauly. Where some critics noted that Ecopath was too simplistic to realistically model the complexity of a coral reef ecosystem, Dr. Pauly saw its simplicity as a strength that would allow it to be applied to many ecosystems around the world and he began doing just that. It wasn't long before Dr. Pauly had the good fortune to develop a collaboration with the two co-authors of this book, Dr. Villy Christensen and Dr. Carl Walters, to build a much more user-friendly and temporally and spatially dynamic version of the model that became known as Ecopath with Ecosim (EwE). Over the years, Dr. Christensen has been a leader in working to continually improve EwE and develop a global community of EwE users and developers. Due to his efforts and the utility of this modeling approach, the number of scientists that have received training in EwE exceeds 3200 coming from more than 100 countries – it's truly a global initiative.

Not long ago, I had the opportunity to explore the ecosystem and fisheries impacts from both climate change and fishing strategy by driving an EwE model for the central Pacific with the output of an earth system model. It was very impressive to see what a versatile and powerful tool EwE has become. This book, written by lead EwE developers, is a timely and comprehensive presentation of this versatile modeling approach and its many applications for the benefit of current and future ecologists, resource managers, and policy makers.

Jeffrey Polovina, Ph.D.

Kailua, HI 96734

Foreword

Sheila JJ Heymans

Ecopath with Ecosim (EwE) has been a constant in my life for nearly 30 years. I first became aware of it as a PhD student in South Africa, where I was in the depths of building a model using the labour- and data-intensive software, NETWRK, an ecological network analysis (ENA) tool written in FORTRAN by Bob Ulanowicz. I was working on a model of the northern Benguela current, and my PhD supervisor, Dan Baird, came back from the USA with this new software, Ecopath – on two discs with no manual. Luckily NETWRK had been included into the Ecopath software by Villy Christensen. Long story short, I spent quite some time trying to get the same results from Ecopath as we got in NETWRK, and in that process realized that Ecopath is much more forgiving of data gaps. It is also much more logical to an ecologist, which is why it has been so successful over the past four decades.

The theories behind network analysis and Ecopath has influenced my view of the world in everything from understanding the behaviour of seagulls on garbage day, to politics: human systems become brittle when you reduce diversity, just like ecosystems do, so reduced redundancy in political viewpoints creates two party systems that flip-flops from one extreme to the other, or in the worse-case one party systems that become dictatorships – just like lack of predators changes the dynamics of coral reefs. Similarly, the foraging arena theory explains why the seagulls in Oostende will always peck at my garbage bags until the municipality starts using garbage cans, reducing the vulnerability of the garbage to the predation by seagulls, similar to the small fish under Carl's boat in Figure 6 of the On Modelling chapter were not predated on by the larger fish, or how the planktivores that stay closer to the corals avoid predation in Figure 7, and explained more mathematically by Figure 4 in the Density Dependence chapter. So, if you study the networks of fisheries, or the networks of fishers, as you will be able to do after reading this book and following the tutorials, you too will be able to understand some of the crazy things happening in the world.

Seriously though, to address the very real problems that we face both in the Ocean and on land, we need all the scientists and tools we can get, and EwE is one of the few tools that translate easily to people with limited programming skills. The software has been taken up by policy makers in the USA, Australia and South Africa more quickly than in Europe, but even this bastion of academic conservatism is realizing that we need all the tools we can get to help with the large problems we have created. In the past decade EwE has been used in at least 15 large scale projects in Europe, through both National, Regional and European funding. This has created a suite of models that is ready to be plugged into the EU new Destination Earth architecture: the European Digital Twin of the Ocean (EDITO). As part of the EU's Mission to Restore our Ocean and Waters, EDITO must make Ocean knowledge available to all and be usable to address what-if questions asked by policy makers not just in the Ocean but also on land, where most of the problems originate. EwE is the most comprehensive and understandable tool to achieve those aims, and these models will be critical to ensure that EDITO achieve those aims.

The EU is also currently reviewing its Marine Strategy Framework Directive, which needs better indicators for food web descriptors, and under the EU's Biodiversity Strategy for 2030 should dovetail more closely with other legislation such as the Water Framework Directive, Nitrates Directive, the Maritime Spatial Planning Directive and the Common Fisheries Policy. I do not know of any other tool that will be able to address this better than an EwE-enabled EDITO.

Sheila JJ Heymans

Executive Director, European Marine Board

I

Ecosystem Modelling

1.

On modelling and making predictions

Villy Christensen

Figure 1. Raymond Lindeman (1915-1942).

Food web analyses (and with them ecological networks), as we know them, dates back to the pioneering studies of Raymond Lindeman around 1940 (Figure 1). He studied Cedar Creek Bog in Minnesota and made a detailed model of nutrient cycling expressed as energy flows[1] (Lindeman 1942). For this, he used thermodynamic principles to evaluate and understand ecosystem functioning, and through this he established the field of trophic dynamics. The study of energy flows and concepts he introduced, such as food chains, food webs, ecological transfer efficiency, and energy pyramids, now provides core elements of community and ecosystem ecology.

Lindeman received a fellowship to work with G. Evelyn Hutchinson at Yale University, managed to publish his PhD studies on Cedar Creek Bog though ill, but unfortunately died soon after, only 27 years old. He was a brilliant mind, and we can only guess how he would have shaped our research world had his days been more numerous.

Lindeman's studies, however, inspired research for decades to follow. Most notably, the International Biological Program (IBP), a major international initiative that during 1964-1974 conducted studies of biological productivity in ecosystems throughout the world. Incidentally, this was also where I first participated in ecological research as a first-year student joining the tail end of the study, sampling fish in a lake in Denmark.

Figure 2. Study sites of the International Biological Program (IBP).

The IBP was mainly descriptive in its nature, and had numerous modelling activities including some dynamic ecosystem modelling – a topic to which we return later. A lasting legacy of the IBP was that it brought focus to ecosystem research. There were also numerous follow-up studies to the IBP. Methodologies had been developed and coordinated through the IBP, and many researchers had been introduced to the field. The time had come for ecosystem research.

Among the follow-up studies was an extensive five-year study conducted around 1980, of the French Frigate Shoals in the Northwestern Hawaiian Islands. Researchers quantified energy flows and biomasses ranging from plankton through to marine mammals, and over the five years gathered an impressive amount of data. Realizing the need to make sense of the mountain of data, NOAA hired a newly graduated oceanographer, Jeff Polovina, to construct an ecosystem model of the French Frigate Shoals.

At this time there were two major activities on ecosystem modelling with a fisheries perspective. Taivo Laevastu and colleagues at the NMFS Alaska Fisheries Science Centre worked on a complex multispecies model of the Bering Sea[2] (Laevastu and Larkins 1981) while K.P. Andersen and Erik Ursin, at the Charlottenlund Castle, Danish Institute for Fisheries and Marine Research, were constructing an equally complex model of the North Sea[3]. Polovina evaluated these modelling efforts and realized the impossibility of constructing species-based dynamic models for biologically diverse areas such as a tropical coral reef ecosystem. From the Laevastu model, he adopted the principle of mass-balance, and used this to construct a simple ecological accounting system, which he termed Ecopath.[4]

Mass-balance here means that energy input has to balance energy output (including storage) for each species (or functional group) that is being modeled. If we can mass-balance one species, we can balance the whole ecosystem. For this, we use information about how much food predators require to compare to how much production is available from their prey. It has to match. And what is important, this adds constraints to the modelling. Adding constraints is fundamental for all modelling, and is one reason that mass-balance modelling has shown successful. Along with the ease of application this, in 2009 led to the Ecopath modelling approach, (see Figure 3) being recognized by NOAA as one of the ten biggest scientific breakthroughs in the organization's 200-year history.

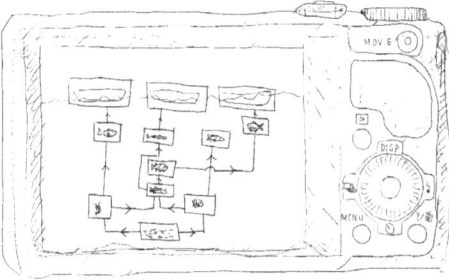

Figure 3. The basic Ecopath model creates a snapshot of an ecosystem at a given point in time: who eats who and how much? Mass balance links predator and prey: there has to be enough food for the predators

I have worked with development of the Ecopath with Ecosim (EwE) approach and software for more than three decades, starting off with Daniel Pauly in the Philippines[5]. Daniel had the idea of merging Polovina's Ecopath model with ecological network analysis such as developed by Robert Ulanowicz[6] and others. Finding out how and seeing it through became my PhD work, which was focused on network analysis of trophic interactions based on meta-analysis of aquatic ecosystems.

Figure 4. Eugene P. Odum (1913-2002).

From this work, let me highlight ecosystem development. One of the greatest ecologists of all times, EP Odum (Figure 4) described a set of ecosystem attributes, and how these would change as ecosystems develop[7]. I quantified most of Odum's 24 attributes based on some forty Ecopath models, and ranked the models based on maturity[8]. It worked really well, and since then a number of colleagues have repeated the analysis with the same result. We can rank ecosystems.

It's typical indicator work. You set a number of criteria, extract the numbers, and out comes a ranking. But what attributes and indicators should we use and how do we obtain the overall ranking? I was really fascinated by this during my PhD: that one could extract a few indicators from food webs and use that to characterize the state of ecosystems.

There are, however, very many indicators and properties in ecological network analysis

– you can get the impression that any ecologist doing research in the field in order to be noticed must develop their own way to capture the essence of ecosystems. This, aggravated by very little attempt at evaluating methods and approaches across studies, seems to characterize the field: consensus building has not been an integral part of the development. The big challenge after half a century of ecological network analysis is still to explain what the seemingly endless suite of indicators tells us.

Yet I do not intend to compare network analysis to the *"Emperor's New Clothes"* (Figure 5) – though it is a challenge to interpret the many concepts and indicators. I have worked enough with network analysis to see clear patterns, some of which are consistent and rather straightforward to explain, while others are much more elusive. As an example of where I still have unfulfilled expectations of network analysis, let me point to identification of critical species in ecosystems – the canaries in the coal mine, and as part of this, what makes an ecosystem vulnerable to perturbations?

Figure 5. Food web representations can be beautiful, but what do they tell us?

I come to think of the *Hitchhiker's Guide to the Galaxy*, especially the third of five volumes in the trilogy[9] If you don't remember it: our planet was really a giant super computer operated by mice. It tolled away for millions of years to answer the biggest and most fundamental question about *Life, The Universe and Everything*. Eventually the answer came: 42, but by then no one remembered the question. I've often been in that situation with network analysis and indicators: It gives the answer, but what was the question? What do the indicators tell us? How do we interpret them? And importantly, can we use this for making prediction?

Making predictions and evaluating "what if" questions remain elusive, however, as ecological network analysis has demonstrated very little predictive capabilities, such as we are craving for fisheries management. Rather, network analysis tends to be static, almost without exception – it's the study and interpretation of snapshots such as mentioned earlier.

Dynamic considerations have, however, entered from a different route. There was a productivity sub-group of IBP that focused on modelling, including dynamic modelling of ecosystems. For this, they created a new field in ecology, systems analysis, and recruited a cohort of bright, quantitative young scientists that used the emerging computers to make models and analysis never imagined before.

In essence, what they did was turning the snapshot from the static food web studies into the movie version. And somehow a movie is less open to interpretations than a photo: it adds constraints. But the modelling had problems. All predator-prey modelling is in essence built on Lotka-Volterra dynamics. This means that the consumption by predators is estimated from the product of the number of predators, the number of prey, and a search rate. More predators more consumption; more prey more consumption. Behind this is a thermodynamic principle called mass-action, and this works absolutely fine when mixing reagents and wanting to predict the products. There are, however, problems when using it in ecology.

The systems analysts in the IBP found that their dynamic models were unstable, and commonly experienced cycles and model self-simplification. Cycles are fine when modelling for instance snowshoe hare – lynx interactions in boreal systems[10], but they are not regular features of more diverse ecosystems. What presented a bigger problem was self-simplification: Lotka-Volterra models are inherently unstable, and it is not possible to maintain ecologically similar groups in models with top-down, mass-action control. The poorer competitors will die out. This was a problem that marred the modelling of ecosystems, and eventually most or all of the IBP modellers left the field to pursue other avenues.

Figure 6. The birth of the foraging arena theory.

One of the bright young fellows in the IBP was my colleague Carl Walters. He had struggled to make ecosystem models behave and given up[11]. Then one day in the early 90s he was out fishing on a lake in BC with his 9-year-old son, Will. When you fish with Carl you don't often catch anything, so Will got bored, looked over the side, and saw a lot of nice big *Daphnia* in the water (Figure 6). He asked: "Why don't the fish eat them all, Dad?"

Carl went on to give the obvious explanation, one that any fish biologist could have given. *"We are fishing for big trout, they are out here in the open and deep part of the lake. The small trout hide along the shore where the big ones don't come, and it's the small ones that eat Daphnia. If the small trout come out here, they will be eaten by the big ones*". A simple straightforward explanation, and only afterwards did the profound implications of the reply dawn on him.

The fundamental aspect missing in predator-prey modelling was behavior. Organisms are not randomly moving particles as thermodynamics and mass-action terms tell us. Think of a coral reef with its swarms of planktivores. The small stay close to the safety of the reef, the larger stray a bit further away, but only a safe distance. The moment a roaming piscivore, such as a barracuda, comes patrolling by, they all take cover.

The implication of this is that the prey concentration the piscivores sees is different from the total planktivore abundance, just like the plankton concentration we may measure with nets around the reef is different from what the planktivores actually experience when their foraging is restricted to the immediate safe surroundings of the reef. It takes three to tango: the planktivore (dancer one) restricts its activities in response to the piscivore (dancer two), and this in turn restricts its own access to plankton (dancer three)[12].

From a modelling perspective, Walters developed an elegant way of adding behavior to the predator-prey modelling through the foraging arena theory[13]. Organisms change between two behavioral states, being available or unavailable for predation, and including this only calls for adding one additional parameter to the Lotka-Volterra equation, a behavioral exchange coefficient (that relates to carrying capacity).

Figure 7. Coral reef representation of the foraging arena – the fish are planktivores and stay close to the reef, alert and ready to dive for cover.

One small step of logic, but a giant step for modelling – suddenly the ecosystem models started behaving. Where it had been almost impossible to get models to maintain diversity, incorporation of the foraging arena considerations opened for replicating the known history of ecosystems. This started in earnest a decade ago when fitting ecosystem modelling to time series data started proliferating, and we have since witness a virtual explosion of case studies to the effect that there probable now are more than a hundred of the kind (Figure 8).

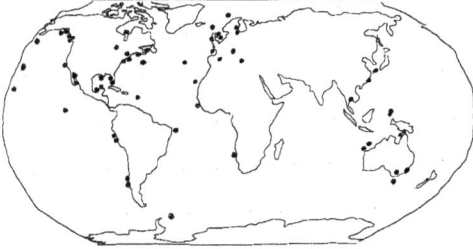

Figure 8. Case studies where Ecosim models have been fitted to time series data. The figure was made in 2011, and the number has by now probably tripled or more.

The case studies are based on the Ecosim module of the Ecopath with Ecosim (EwE) approach and software[14], and we have drawn a number of lessons from them[15], including what you'll read in this textbook. As a rule, to explain historic changes in ecosystems we have to consider,

1. Food web effects,

2. Environmental change, and

3. Human impact, (see Figure 9).

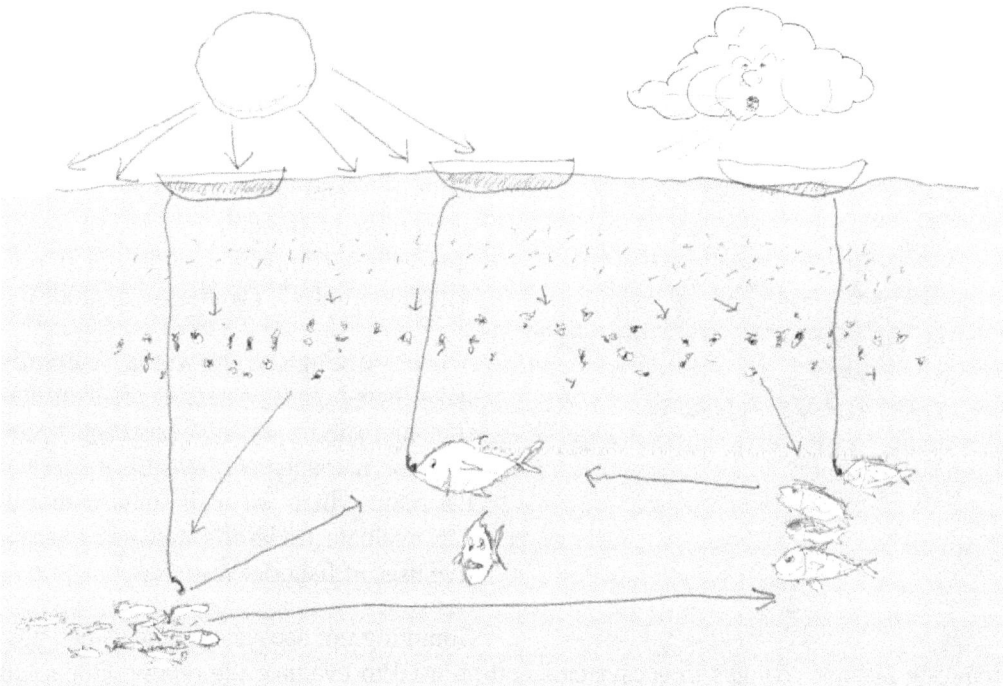

Figure 9. Replicating the history of ecosystems calls for inclusion of food web, environmental, and human impact.

An implication of this is that environmental productivity patterns can be identified throughout the food web. There are variable time delays linked to turnover rates and food web constellations, but we can see environmental signals propagate through the food web. We have also seen evidence that environmental productivity can be amplified through the food web. The biological explanation for this may be that more food results in more excess beyond maintenance, freeing resources to be allocated to growth and reproduction.

Fitting-wise, the models tend to work well for species or groups with strong fisheries impacts, i.e. we generally find good agreements with single-species assessment models. Where there are divergences, they can often be explained from model assumptions related to food web effects. It is also clear that while trends for some species can be explained, there can be others for which the models are unable to offer insight – often because we have no reliable information about what the important drivers of change may be for such species. There is, however, nothing to indicate that such model failures have implications for the overall model fit – here one rotten apple doesn't spoil the bunch.

Figure 10. The butler did it: humans are the usual suspects when evaluating fish population trends, but ecosystem models can now be used to evaluate the relative contribution of food web, environmental, and human impact.

We see impacts of changes in predator abundance on forage species (prey release), and in some cases the opposite effect; where prey abundance impacts predators. Also, there are cases where fisheries seemingly outcompete predators as increased fishing mortality on a forage species is accompanied by decline in predation mortality[16] (Walters *et al.* 2008).

Where ecological networks currently have their biggest potential for contribution in fisheries is for evaluating trade-offs for management. We have reached the point where we with some authority can evaluate trade-offs between alternative uses of fisheries resources[17].

Summing up, ecosystem models can now replicate historic changes in ecosystems and be used to evaluate the relative impact of fisheries, food web dynamics, and environmental change (Figure 10), and notably use this to evaluate trade-offs. With models that behave well enough to replicate the past, we can start thinking of using them to predict the future, to ask "what-if" questions.

Figure 11. Will there be seafood and healthy oceans for future generations to enjoy?

The key question we have to ask is "will there be seafood and healthy oceans for future generations to enjoy?" (Figure 11). To answer the question, we have to make predictions. There will be uncertainty and unexpected events, but we need to provide guidelines and options – to ensure that there will be seafood for future generations. What choices must we make for this?

Given that the seafood market is an international one, it is a global question, and we have to tackle the question through modelling scaled accordingly. There is, however, no tradition for global modelling in fisheries, and while the Intergovernmental Panel for Climate Change, IPCC, has done the necessary job on predicting how our physical environment will be impacted by climate change, it is only in recent years that the consequences of climate change on life on earth has gained attention.[18]

Figure 12. When making predictions, expect the unexpected. The vampire in the basement will bite you.

The Intergovernmental Science-Policy Platform on Biodiversity and Ecosystem Services (IPBES) has taken on this task, and from an aquatic modelling perspective this work is supported by the Fisheries and Marine Ecosystem Model Intercomparison project, Fish-MIP), which works to develop a global ocean-modelling framework that incorporates modelling of the physical environment, of lower and higher trophic levels, and of human activities including governance. Fish-MIP provides a framework with alternative modelling components in order to consider uncertainty through an ensemble approach, following the lead for how the IPCC has tackled global environmental modelling.

Uncertainty indeed has to be a major factor in making predictions. While ecosystem models now offer some predictive capabilities for evaluating major human impacts and making predictions, we cannot make beautiful orchestrated symphonies or detailed predictions, and we will never be able to do that for complex ecosystems. There are notably two factors that prevent this. One is Walter's "vampires in the basement" (Figure 12), the other is incomplete knowledge of how systems will react to management interventions.

We must expect the unexpected; there will be events we cannot predict. Invasive species is a case in point, and more generally, behavioral responses in ecosystems are no more predictive than they are for human systems. Let me illustrate with an example; seals have been increasing in the Strait of Georgia since culling ceased in the 1970s. For about 30 years thereafter, mammal-eating transient killer whales were rarely observed in the Strait. Then one summer, a small pod came in and found plenty of prey – the next summer the whale watching boats counted a hundred transient killer whales coming in, and transients have been regular visitors since then. From a modelling perspective, such behavioural events are unpredictable, and they have repercussions through the ecosystems.

Figure 13. Monitor, experiment, and adapt. The fundamental aspects of adaptive management rely on modelling as the guiding factor.

There is also considerable uncertainty about how ecosystems will react to many management interventions, especially where our knowledge about drivers and impact is very incomplete. Our best option wherever this is the case is represented by adaptive management with carefully planned monitoring, experimentation, and adaptation[19]. Modelling is an integral part of this, needed to guide the entire process and limit the risk of making bad, preventable mistakes.

Figure 14. Alice: *"Would you tell me, please, which way I ought to go from here?"* **Cheshire Cat:** *"That depends a good deal on where you want to get to".* **Policy makers need to set clear objectives for management, and scientists need to evaluate alternative options for managers.**

So, though we cannot make detailed predictions for how ecosystems will develop, we as a society need to carefully choose what direction to take and we need to avoid the preventable mistakes. For this, it is crucial that fisheries policy makers and managers set clear objectives for management, and that fishery scientists in turn define and evaluate alternative policy options (Figure 14). We need to manage our ecosystems with a strong commitment to moving in a sustainable direction if there indeed is to be seafood and healthy oceans for future generations to enjoy.

Acknowledgements: With special thanks to Dalai Felinto for the original artworks. To Carl Walters for discussions that helped shape this contribution and for the many years of work that went before it. Also to Buzz Holling, Steve Carpenter, Eddie Carmack, and Daniel Pauly for discussions and inspiration, to Rhys Bang Williams for representing the future generations, and to Bill Fisher and the American Fisheries Society for the opportunity to address the 142nd Annual Meeting with the opening lecture *"Ecological Networks in Fisheries"* on which this chapter is based.

Attribution: The chapter was adapted from Christensen, V. 2013. Ecological networks in fisheries: predicting the future? Fisheries, 38(2): 76-82 with License Number 5642170043159 from John Wiley and Sons. https://doi.org/10.1080/03632415.2013.757987. Rather than citing this chapter, please cite the source.

Notes

1. Lindeman, R.L. 1942. The trophic-dynamic aspect of ecology. Ecology 23, 399–418.

2. Laevastu, T. and Larkins, H.A. 1981. *Marine fisheries ecosystem: its quantitative evaluation and management.* Fishing News Books, Farnham, England.

3. Andersen, K.P. and Ursin, E. 1977. A multispecies extension to the Beverton and Holt theory of fishing, with accounts of phosphorus circulation and primary production. Meddelelser fra Danmarks Fiskeri og Havundersøgelser 7, 319–435.

4. Polovina, J.J. (1984) Model of a coral reef ecosystem. Coral Reefs 3, 1–11

5. Christensen, V. and Pauly, D. 1992. ECOPATH II — a software for balancing steady-state ecosystem models and calculating network characteristics. Ecological Modelling 61, 169–185.

6. Ulanowicz, R.E. 1986. *Growth and Development: Ecosystem Phenomenology.* Springer Verlag (reprinted by iUniverse, 2000), New York.

7. Odum, E.P. 1969. The strategy of ecosystem development. Science (New York, N.Y.) 104, 262–270.

8. Christensen, V. 1995. Ecosystem maturity - towards quantification. Ecological Modelling 77, 3–32.

9. Adams, D. 1982. *Life, The Universe and Everything.* Harmony Books, New York.

10. Krebs, C.J., Boonstra, R., Boutin, S. and Sinclair, A.R.E. 2001. What drives the 10-year cycle of snowshoe hares? Bioscience 51, 25–35

11. Hilborn, R. and Walters, C.J. 1992. *Quantitative Fisheries Stock Assessment: Choice, Dynamics, and Uncertainty.* Chapman and Hall.

12. Walters, C.J. and Martell, S.J.D. 2004. *Fisheries Ecology and Management.* Princeton University Press, Princeton

13. Ahrens, R.N.M., Walters, C.J. and Christensen, V. 2012. Foraging arena theory. Fish and Fisheries 13, 41–59.

14. Christensen, V. and Walters, C.J. 2004. Ecopath with Ecosim: methods, capabilities and limitations. Ecological Modelling 172, 109–139

15. Christensen, V. and Walters, C.J. 2011. Progress in the use of ecosystem modelling for fisheries management. In: *Ecosystem Approaches to Fisheries: A Global Perspective.* (eds V. Christensen and J.L. Maclean). Cambridge University Press, Cambridge, pp 189–205.

16. Walters, C., Martell, S.J.D., Christensen, V. and Mahmoudi, B. 2008. An Ecosim model for exploring ecosystem management options for the Gulf of Mexico: implications of including multistanza life history models for policy predictions. Bulletin of Marine Science 83, 251–271.

17. e.g., Christensen, V. and Walters, C.J. 2004. Trade-offs in ecosystem-scale optimization of fisheries management policies. Bulletin of Marine Science 74, 549–562.

18. Schmitz, O.J., Raymond, P.A., Estes, J.A., Kurz, W.A., Holtgrieve, G.W., Ritchie, M.E., Schindler, D.E., Spivak, A.C., Wilson, R.W., Bradford, M.A., Christensen, V., Deegan, L., Smetacek, V., Vanni, M.J., Wilmers, C.C., 2014. Animating the carbon cycle. Ecosystems 344–359. https://doi.org/10.1007/s10021-013-9715-7

19. C. J. Walters, 1986. *Adaptive Management of Renewable Resources,* MacMillan, New York, Reprint 2001.

2.

Modelling predator-prey interactions

On the path to ecosystem-based management: species interactions

Let there be no doubt, single species assessment is a necessary factor for management of fisheries, notably for tactical management. How do we manage this species in this bay this year? What we need to ask, however, is if it is sufficient?

Views on this question go back a long time as expressed by the two pioneers that more than any established fisheries science as a quantitative discipline, Ray Beverton and Sidney Holt. In their 1957 Magnus Opus, "On the Dynamics of Exploited Fish Populations"[1] they wrote (p.24):

"Elton (1949) has suggested that the goal of ecological survey is '…to discover the main dynamic relations between populations living in an area'. This is a generalization of what is now perhaps the central problem of fisheries research: the investigation not merely of the reactions of particular populations to fishing, but also of interactions between them and the extent to which it is possible and practicable to derive laws describing the behavior of the community from those concerning the properties of component populations"

Ray Beverton and Sidney Holt in their book set the vast part of the agenda that fisheries scientists have worked on ever since. So is the case for species interactions as the quote above illustrates. If the assessments are short-term, as single species assessments tend to be, we can get by assuming "business as usual", but when we move away from the initial state, i.e. when we address questions at the ecosystem-level we have no choice.

"Fish eat fish", Erik Ursin – who along with K.P. Andersen created one of the first end-to-end ecosystem models[2] – often said. And yes, fish eat fish, and that has implications for management of living resources in aquatic environments. If we are to successfully manage ecosystems, then species interactions is part of the foundation.

The foundation for this was laid a century ago, when Lotka[3] and Volterra[45] both and independently formulated a theory for predator-prey interactions. Models based on these sources are called Lotka-Volterra models and they are in essence the foundation for all predator-prey models, including the dynamic models in EwE.

The basis is that with no resource limitations and no predators, prey populations (N) will change over time with an exponential growth rate (r),

$$dN/dt = rN \qquad (1)$$

and predator populations (P) will decrease with a mortality rate (m, due to e.g., starvation or old age),

$$dP/dt = -mP \tag{2}$$

In a simple predator-prey system with no resource limitations for prey, the equations can be coupled. For the prey population, we describe the change over time with the differential equation,

$$dN/dt = rN - aNP \tag{3}$$

where the factor a is called the search rate. For the predator,

$$dP/dt = gaNP - mP \tag{4}$$

where g is the growth efficiency with which the predator converts consumption to production.

If you examine these equations carefully, you'll notice that the predator's consumption (Q) is calculated as

$$Q = aNP \tag{5}$$

which means that more predators (P) lead to more predation, and more prey (N) means more predation – the consumption is the product of the two and the search rate constant. Systems modelled with such assumptions are unstable, initially the predator population may grow if there are plenty of prey around, but as it grows, the prey population gets more impacted and at some point the prey will collapse. The slower-growing predators will survive for a while, but in a simple two species systems, they will eventually collapse as well. That in turn releases predation mortality from the minuscule prey populations, which will have great conditions and start growing exponentially – after which history repeats itself, the predator will increase, the prey collapse, the predator collapse. The system becomes cyclic and unstable.

Volterra (1928) summarized the properties of predator-prey interactions in three "laws"

Law of the periodical cycle: The fluctuations of two species populations, where one feeds on the other, are periodic, and the period depends entirely on the coefficients of growth (r) and decay (m) and initial conditions (No and Po).

Law of the conservation of averages: The averages of population numbers of the two species remain constant and independent of the initial values of both populations if and only if the coefficients of growth and decay and the conditions of predation (prey losses, predator gains, i.e. the four coefficients *r,a,e,m*) remain constant.

Law of perturbation of averages: If individuals of both species are removed, (e.g., by predation or fishery) uniformly and in proportion with their total population, the average population of the prey increases, while the average population of the predator decreases. On the other hand, increased protection of the prey species will lead to growth of both populations.

The cyclic and unstable nature of Lotka-Volterra systems is not what we see in most ecosystems, and for that reason there have been numerous modifications proposed, notably after C.S. Holling[6] added predator handling time (h) through his disk equation where the consumption by a predator is estimated as,

$$Q = aNP/(1 + hN) \tag{6}$$

recognizing that the time a predator spends handling a prey it will not be searching for new prey. That limits the predator consumption rate at high prey densities – and leads immediately to unstable dynamics (exploding cycles) in the predictions unless some other factor(s) are included in the model to limit variation in prey and/or predator abundances and predation rates.

Figure 1. Population growth as a function of population size for the logistic (Verhulst) model. Carrying capacity for the population is set at 2.

Lotka-Volterra models can be defined without (as above) or with resource limitation, i.e. carrying capacity (*K*). The "standard" way of implanting resource limitation is to express prey population change using the logistic equation for population growth (Verhulst),

$$dN/dt = rN(1 - N/K) \tag{7}$$

Foraging arena

The foraging arena theory was developed by Carl Walters, and serves as the foundation for the dynamic modules of EwE, i.e. of Ecosim and Ecospace. The basic assumption in foraging arena theory is that spatial and temporal restrictions in predator and prey activity cause partitioning of prey populations into vulnerable and invulnerable population components, such that predation rates are dependent on (and limited by) exchange rates between these prey components[7].

Foraging arena models (such as Ecosim[8]) are based on Lotka-Volterra modelling but the interaction terms only include the vulnerable part (V) of the total prey population (N). So, where we for Lotka-Volterra models have the predator consumption (Q) estimated from,

$$Q = aNP \tag{8}$$

the similar equation for foraging arena models is,

$$Q = aVP \tag{9}$$

Further, the prey exchange between vulnerable and in vulnerable states can be described with the rate equation,

$$dV/dt = v(N - V) - v' - aVP \tag{10}$$

from which V is predicted to the moving equilibrium (for N and V), setting dV/dt to 0,

$$V = vN/(v + v' + aP) \tag{11}$$

Changes in the predator and prey populations over time can then be predicted from the Lotka-Volterra model equations, substituting the total prey populations (N) with the vulnerable population (V).

Associated tutorial

There is a tutorial to accompany this section in the next chapter of the web version of this book. You can either develop a Lotta-Volterra model based on the equations above or use the R-code that is included in the tutorial (or do both).

Quiz

An interactive H5P element has been excluded from this version of the text. You can view it online here:
https://pressbooks.bccampus.ca/ewemodel/?p=394#h5p-1

Notes

1. Beverton, R.J.H. and Holt, S.J. 1957. On the dynamics of exploited fish populations. Fisheries Investigations, 19, 1-533.

2. Andersen, K.P. and Ursin, E. 1978. A multispecies extension to the Beverton and Holt theory of fishing, with accounts of phosphorus circulation and primary production. Meddelelser fra Danmarks Fiskeri- og Havundersøgelser, 7, 319-435.

3. Lotka, A.J. 1925. Elements of Physical Biology. Williams and Wilkins, Baltimore

4. Volterra, V. 1926. "Variazioni e fluttuazioni del numero d'individui in specie animali con-viventi". Mem. Acad. Lincei Roma. 2: 31–113.

5. Volterra, V. 1928. Variations and fluctuations of the number of individuals in animal species living together. J. Cons. int. Explor. Mer 3(1): 3–51.

6. Holling, C. S. 1959. The components of predation as revealed by a study of small mammal predation of the European pine sawfly. Can. Ent. 91: 293–320.

7. Ahrens, R.N.M., Walters, C.J., Christensen, V., 2012. Foraging arena theory. Fish Fish. 13, 41–59.

8. Walters, C., Christensen, V., and Pauly, D. 1997. Structuring dynamic models of exploited ecosystems from trophic mass- balance assessments. Reviews in Fish Biology and Fisheries, 7(2): 139–172.

3.

Your research question?

The simplest model that can address your research question is the best model

So you want to build an ecosystem model. It should be very capable, it should be able to address any policy, management, environmental, population dynamic or ecological question about the resources in your ecosystem. If so, it has to be complex, it has to be end-to-end, include everything important from nutrients over whales to how society works. Right?

No, that thinking leads to Frankenstein models, models that take on a life of their own where no-one really knows why they behave the way they do. Instead the rule is: the simplest model that can address a given question is the best model to use. Best, because it's the model with most predictive power[1].

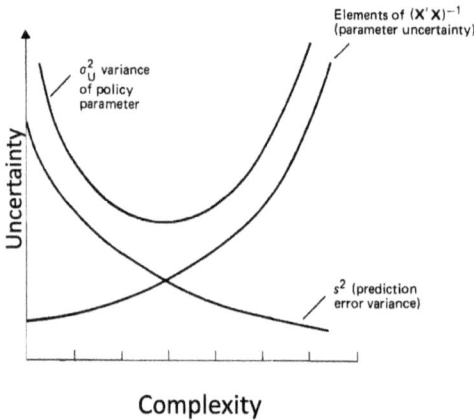

Figure 1. Uncertainty about a policy parameters U (as expressed through its variance σ_U^2 is likely to be minimized at an intermediate complexity. Modified from Walters (1986[2]).

That rule also holds for working with EwE. Don't spend years building a complex ecosystem model in order to address any question that may arise. Instead, close doors as you move ahead, build something simpler, as simple as you can get away with – you can always add details and see if it makes a difference. The model building process is a continuum you may never finish, recognize that and focus your effort. Make it a simple as you can in order to address the one thing that should be your guiding star: What is your research question?

Descriptive or predictive model? If descriptive you want to be inclusive, if predictive, be selective. See Defining the ecosystem.

Keeping it simple has several advantages, faster development and run times are just part of the picture. Equally as important is that keeping it as simple as possible makes it much more realistic to develop alternative models to address the research question that guides the work. It also makes it easier for you, the modeller to grasp what your model is doing and why.

Proponents of this principle have long advocated the use of Minimum Realistic Models[3] and Models of Intermediate Complexity for Ecosystem assessments[4], and this is indeed an approach that can go hand-in-hand with EwE model development and application.

> Models are not like religion
> – you can have more than one
> – and you shouldn't believe them

Part of this is that models are not like religion, you can have more than one (and you shouldn't believe them). So, if you find that a model can answer your question, should you believe the answer? No, evaluate how robust your findings are, don't take them at face value. Examine your model and explore what it takes to make the answer go away. That doesn't mean they are always wrong, but if you keep your model relatively simple, you can use alternative model formulation to explore the same or different questions. Consider and test if there are alternative hypotheses (and with it model formulations) that may be used to address the questions that drive your research.

Consider where we would be with climate change, if the IPCC had decided years ago to build one-model-that-rules-them-all to evaluate the mechanisms and impacts of climate change. We would still be arguing if what we are experiencing is climate change or natural variations. The approach that IPCC have taken by using an ensemble model approach represents the best way forward. IPBES has picked it up, and are using an ensemble approach to evaluate among other the impact of climate on fisheries and marine ecosystem through the Fish-MIP model intercomparison approach – of which EwE models are an integral part.

> If you are thinking of a management intervention, and you can't make it work in a simple model, what's the chance it may work in reality?
>
> Keith Sainsbury

There may be an apparent conflict between keeping it as simple as possible and using an ensemble approach of perhaps more complex models, but that conflict is indeed only apparent.[5]

Holden et al.[6] provides a cost-benefit analysis of ecosystem modelling to support fisheries management. As part of this they evaluate not just benefit but model costs as well for models ranging from single-species models to complex ecosystem models. They found that costs varied by two orders of magnitude with cost increasing with model complexity. There are trade-offs and no "correct" way or "best" model. What matters is what is needed to answer the research/policy questions. That has to be defined from the very onset.

Notes

1. Walters, C.J. 1986. Adaptive Management of Renewable Resources. MacMillan

2. Walters. 1986. op. cit.

3. D. S. Butterworth, É. E. Plagányi. 2004. A brief introduction to some approaches to multi-species/ecosystem modelling in the context of their possible application in the management of South African fisheries, African Journal of Marine Science, 26:1, 53-61

4. Plagányi, É., Punt, A., Hillary, R., Morello, E., Thebaud, O., Hutton, T., Pillans, R., Thorson, J., Fulton, E.A., Smith, A.D.T., Smith, F., Bayliss, P., Haywood, M., Lyne, V., Rothlisberg, P. 2014. Multi-species fisheries management and conservation: tactical applications using models of intermediate complexity. Fish Fisheries 15:1-22. https://doi.org/10.1111/j.1467-2979.2012.00488.x

5. J. Elder. Chapter 16 - The apparent paradox of complexity in ensemble modeling R. Nisbet, G. Miner, K. Yale (Eds.), Handbook of Statistical Analysis and Data Mining Applications (Second edition), Academic Press, Boston (2018), pp. 705-718, https://doi.org/10.1016/B978-0-12-416632-5.00016-5

6. Holden, M. H., Plagányi, E. E., Fulton, E. A., Campbell, A. B., Janes, R., Lovett, R. A., Wickens, M., Adams, M. P., Botelho, L. L., Dichmont, C. M., Erm, P., Helmstedt, K. J., Heneghan, R. F., Mendiolar, M., Richardson, A. J., Rogers, J. G. D., Saunders, K., & Timms, L. (2024). Cost–benefit analysis of ecosystem modeling to support fisheries management. Journal of Fish Biology, 1–8. https://doi.org/10.1111/jfb.15741

4.

Defining the ecosystem

The ecosystems to be modelled using EwE can be of nearly any kind: the modeller sets the limits. The general rule for descriptive network analysis is to define the system so that the interactions within the system add up to a larger flow than the interactions between it and the adjacent system(s). In practice, this means that the import to and export from a system should not exceed the sum of the transfer between the groups of the system. If necessary, one or more groups originally left outside the system should be included in order to achieve this.

For predictive models, it's a bit simpler or at least different. You formulate your research question, identify the key species and include the groups of importance for the key species. That would typically be the predators and prey of the key species along with fisheries that impacts these groups. Given that this indeed is much simpler, we'll focus the rest of this chapter on descriptive models.

What should you include in your model? There is no clear answer to that, it really depends on what your research or policy question is. For this, there are basically two major categories, descriptive vs. predictive.

For descriptive models, you want to be complete, to include as much of the ecosystem components as practically possible. We've often initially erred on the side of being inclusive, adding groups even if they probably had low biomasses, flows and potential impact. Then as the model took shape and it became clear that certain groups were unimportant, we'd aggregate those groups to make things more manageable. The strength of descriptive models is that they can be used to characterize the form and functioning of ecosystem, including notably for network analysis, e.g. about the ecosystem stage of development[1]. But their ability to address 'what-if' questions is severely limited or non-existent – one can use the mixed trophic impact analysis to give indications for what would happen as a result of a change in abundance or fisheries impact, but it will only be indications, not actual predictions.

To make predictions, we need predictive models, which means dynamic models that can be used to address more targeted research or policy questions than those posed to descriptive models. While descriptive models can be used as the foundation for predictive models, they come with a ballast of long development time, longer run time, and a flurry of detailed output that can make it difficult to actually get to addressing the policy / research questions that should drive the effort.

Instead the best advice is to develop predictive models with what is needed to address the questions at hand. Focus on the target species, add important predators and fisheries, and include key prey groups and lower trophic level groups. Start off with a simple model, get to address the questions, then explore what happens if you add more details. So, don't try to make a model that perhaps someday may be useful for addressing a variety of questions, you'll spend so much time and energy doing that that you may never get to actually addressing the questions that were to drive the research.

Descriptive or predictive?

Let's illustrate the difference between descriptive and predictive studies with an example. First, clearly define your research or policy question. Say your task is to understand the importance of the fisheries of the Azores Islands. For that you would build a model of the EEZ and include the commercially important species along with their predators, prey and production system, and you would include the various Azorean fisheries in the model. Given your task, you would include the important skipjack tuna in your model. But skipjack is a highly migratory species with an ocean-wide distribution, not confined to the Azorean EEZ, so while it's important to include skipjack in your model, the skipjack population area is much wider than the EEZ. How do you handle that?

The simple way is to recognize that your model is restricted to the EEZ. You include skipjack tuna with both immigration and emigration. If in the Ecosim model runs, you increase the tuna fisheries, more skipjack will be caught, fewer skipjack will leave the EEZ, and the following year the same number of skipjack will immigrate to the EEZ. Your model will have skipjack included, it can fish them more or less, but it will not impact the overall skipjack population.

Is that OK?

To answer that you have to consider your research question. If you are indeed describing the fisheries of the Azores Islands, all seems good with this approach. But if what you really want is to evaluate skipjack population dynamics, you need to define a model area that encompasses the skipjack distribution area and all-important fisheries. That would not be a model of the Azores Islands, but a dedicated MICE type model, i.e. a simpler predictive model focused on the key species of relevance for your research question.

Define your model question(s), that's your focus!

For descriptive models

The groups of a system may be (ecologically or taxonomically) related species, single species, or size/age groups, i.e., they must correspond to what is called "functional groups." Using single species as the basic units has clear advantages, especially as one then can use estimated or published consumption and mortality rates without having to average between species. On the other hand, averaging is straightforward and should lead to unbiased estimates if there is information about all the components of the group. The input parameters of the combined groups should simply be the means of the component parameters, weighted by the relative contribution of the species in the group. Often one does not, however, have all the data needed for weighting the means. In such cases, try to aggregate species that have similar sizes, growth and mortality rates, and which have similar diet compositions.

There is a facility in FishBase (www.fishbase.org) that assembles, for any country and many ecosystems, a list of the freshwater and marine fish occurring in different habitat types, and other information useful for Ecopath models (maximum size, growth parameters, diet compositions, etc.)

For tropical applications, grouping of species is always needed: there are simply too many species for a single-species approach to be appropriate for more than a few important populations. It is difficult to provide specific guidelines on how to make the groupings, as this may differ among ecosystems. Generally however, one should consider the whole ecosystem, e.g., for an aquatic model, one or two types of detritus (e.g., one to include mainly marine snow, the other discarded bycatch, if any), phytoplankton, benthic producers, herbivorous and carnivorous zooplankton, micro- and macrobenthos, herbivorous fish, planktivorous fish, predatory fish, etc., and that at least 12 groups are included, including the fishery (any number of fleets/gears), if any. But most important is the personal judgment of what is appropriate for your system.

The recommendation of including at least 12 groups is based on Christensen (1994).[2]

Special consideration needs to be given to the bacteria associated with the detritus. One option, applicable in cases where no special emphasis needs to be given to bacterial biomass, production and respiration, is to disregard the flows associated with these processes, which are, in any case, hard to estimate reliably, and which tend to completely overshadow the other flows in a system. (In such cases, one assumes that the bacteria belong to a different, adjacent ecosystem linked to yours only through detritus export). Alternatively, bacteria can be attached to one or all of the detritus boxes included in a system. To do this, create a "box" for the bacteria, and have them feed on one or several of the detritus boxes. (This is required because detritus, in the Ecopath model is assumed not to respire). Consider, finally, that there is no point including bacteria in your model if nothing feeds on them.

For an overview of the ecosystem concept in ecology, we suggest you consult the book by Golley[3].

Open system problems

For almost every defined ecosystem study area, there will be some species that have life cycles that take them outside the defined area for at least part of each year. Movements (exchange) of biomass across the area boundary can be of two types: dispersal, involving unidirectional movement of organisms to and from sink and source populations outside the study area; and migration, involving regular, repeated movements into and out of the area by the same individuals. These are fundamentally different processes, with very different policy consequences. Dispersal acts as an extra mortality-agent and recruitment-source independent of fisheries and other impacts in the study area, while migration exposes organisms from the study area to particular risks and opportunities for part of the time, without acting as a "permanent" drain or source of those organisms.

Dispersal can be represented in both Ecopath and Ecosim by setting immigration and emigration rates in the Other production form in Ecopath. These rates are used in the Ecopath mass balance and are treated in Ecosim as unidirectional (non-migratory) dispersal rates. True migration is more complex to deal with, and Ecosim will give misleading answers if migration is represented only by immigration/emigration rates from Ecopath.

There are two broad options for dealing with directed migration to and from the Ecopath study area so as to avoid misleading predictions in Ecosim:

- The "diet import" approach: for species that migrate to/from the study area for part of each year, include all fisheries/catches that impact the species, independent of whether these are taken within the study area. In the Diet composition form, set the diet import proportion to the proportion of time spent outside the system, and set remaining diet proportions to the diet proportions while in the system times the proportion of time spent in the system. Using this convention, Ecosim then will allow policy exploration of all fisheries that may impact the migratory species, and will treat the food intake rate (per biomass) as constant over time for the time spent feeding outside the system. Ecopath and Ecosim will "automatically" account for reductions in prey impacts caused by the species for the proportion of time that the species spends feeding in outside areas. Note that the list of fisheries impacting migratory species can involve splitting fleets into "inside" and "outside" fishing components (which can be varied or "managed" separately in Ecosim), to represent possible policy changes in where/when the migratory fish are harvested.

- The "model expansion" approach: If it is considered unrealistic to assume that food consumption rates obtained while outside the system (by migratory species) will remain constant over time, then Ecosim must be provided information on possible changes in food organism populations in those outside areas. That is, the outside areas must be "internalized" as part of the modelled system, by adding functional groups representing the outside food web structure. Often, adding such groups may simply mean replicating the initial Ecopath group structure, with the second set of groups labelled "outside species X" and with diet matrix entries set so that the added groups feed on one another but not on the "inside" groups.

A good modelling tactic is to try both approaches and see whether they give different answers. However, note that the first approach can lead to misleading answers upon entry to Ecospace, if the Ecospace mapped area includes the 'outside' system: in that case, the model will continue to "import" part of the diet and food consumption of migratory species. Thus when the model development plan includes use of Ecospace to represent a larger spatial system, the functional group organization for that larger system needs to be included in the initial Ecopath/Ecosim model definition (approach 2).

It is possible to incorporate migration in Ecospace by defining which groups migrate and where their concentration is by month, see Representing seasonal migration in Ecospace for further information.

Attribution

This chapter is in part adapted from the unpublished EwE User Guide: Christensen V, C Walters, D Pauly, R Forrest. Ecopath with Ecosim. User Guide. November 2008.

Notes

1. Christensen, V., 1995. Ecosystem maturity — towards quantification. Ecological Modelling 77, 3–32. https://doi.org/10.1016/0304-3800(93)E0073-C

2. Christensen, V. 1994. Emergy-based ascendency. Ecological Modelling 72:129-144. https://doi.org/10.1016/0304-3800(94)90148-1

3. Golley, F.B. 1993. A History of the Ecosystem Concept in Ecology: More Than the Sum of the Parts. Yale University Press, New Haven, CT.

II

An Introduction to Ecopath

5.

The energy balance of a box

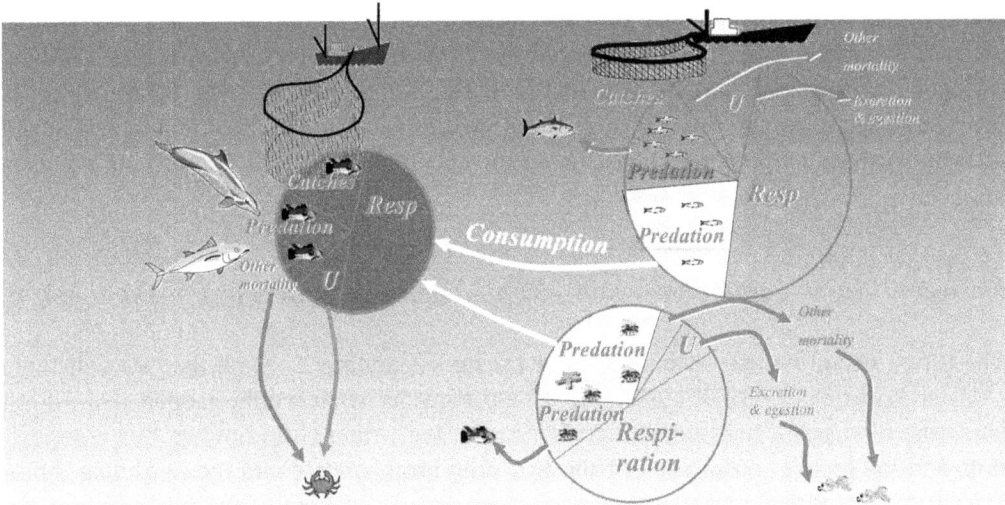

Figure 1. Representation of Ecopath mass-balance (Ecopath "pies"[1]) depicting how the consumption of an intermediary predator can be linked to the production of two prey groups.

Mass balance

Take a close look at Figure 1, it is key to understanding how Ecopath mass balance works. For three of the groups in the system, and intermediary predator, small pelagics and benthos their consumption is represented by a "pie", which size is proportional to the consumption of the group. The predators consumption includes small pelagics and benthos in the proportion dictated by the diet composition of the predator – here that's perhaps 55% for small pelagics and 45% for benthos. Within each of the three groups the consumption is broken into pieces of the pie, using this equation,

If the model currency is a nutrient, there is no respiration, and Eq. 1 becomes *consumption = production + unassimilated part*. In that case, the *unassimilated part =*

consumption – production.

Master Equation 1:

$$\text{Consumption} = \text{production} + \text{respiration} + \text{unassimilated part} \tag{1}$$

where on Figure 1, *Resp* represents respiration and *U* the unassimilated food. This equation is in line with Winberg[2] who defined consumption as the sum of somatic and gonadal growth, metabolic costs and waste products. The main differences are that Winberg (along with many other bioenergeticists) focused on measuring growth, where we focus on estimating losses, and that the Ecopath formulation does not explicitly include gonadal growth. How about predation then? On Figure 1, predation is split into its components, i.e.,

Master Equation 2:

$$\text{Production} = \text{predation mortality} + \text{fishing mortality} + \text{biomass accumulation} + \text{net migration} + \text{other mortality} \tag{2}$$

These two equations are so fundamental for understanding Ecopath that we call them "Master Equations". Check out Figure 1, and consider what would happen if we don't know the biomass of the intermediary predator? We would still know its diet composition, and the production of each of the two prey groups, we could then estimate a biomass for the predator, and see how much they would consume of the two prey groups, and if this was feasible. In that case, production of the prey set constraints for how much the predator potentially can eat. Alternatively, if we didn't know the biomass of one (or both) of the prey groups, the consumption of the predator sets a demand for how much prey there has to be in order to meet the predators' requirements. So, consumption by the predator sets constraints for how prey production. In summary, we use information about the predator consumption to provide constraints for prey production, and information about the prey production to set constraints for predator consumption. The process is called "mass balance", and is conducted throughout the food web, see Figure 2.

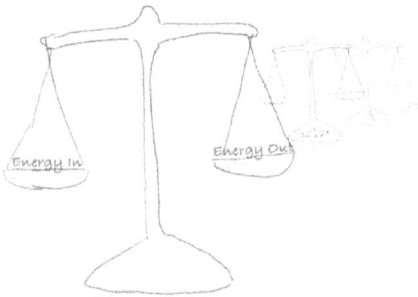

Figure 2. Ecopath is a mass balance model where *energy in* has to equal *energy out* for each groups in the system. *Energy out* for a prey relates to *energy in* for its predators, which links groups in the system and provides constraints for the mass balance.

Does the mass balance process add parameter constraints?

Essington[3] evaluated Ecopath sensitivity to imprecise data inputs, and found that the mass balancing did not have any noticeable effect. The study used nine <u>balanced</u> Ecopath models, added parameter uncertainty and evaluated the degree to which the mass balance could retrieve the "true" parameters values. The study, however, did not recognize that the strength of mass balance is to weed out impossible parameter combinations, so when starting with balanced models those parameter combinations had already been excluded, and minor prediction errors (CV of 0.05 to 0.3) will not make the models sufficiently "unbalanced" compared to models developed from raw data, (which often have conversion errors that the mass balancing is good at pointing to). Our experience is clear, mass balance constrains the parameter space. The mass balance constraint implemented in the two master equations of Ecopath (Eq. 1 and Eq. 2) should not be seen as questionable assumptions, but rather as filters for mutually incompatible estimates of flow. One gathers all possible information about the components of an ecosystem, of their exploitation and interaction and passes them through the mass balance filter of Ecopath. The result is a possible (even plausible) representation of the energetic flows, the biomasses and their utilization. The more information used in the process and the more reliable the information, the more constrained and realistic the outcome will be. The possible representation of state variables and flows is all the Ecopath aims for. Once in the dynamic simulation modules, we can use routines to generate thousands of possible Ecopath models to evaluate impact of uncertainty on policy and research questions.

Parameters

The first Ecopath Master Equation (Eq. 1) can formally be expressed and values estimated from,

$$Q_i = P_i + R_i + U_i \qquad (3)$$

where the parameters are explained in Tables 2 and 3.

Notice that Eq. 3 uses absolute flow rates (t km^{-2} year^{-1}), but in the actual implementation, we estimate the production and consumption as $Q_i = B_i\,(Q/B)_i$ and $P_i = B_i\,(P/B)_i$, respectively. The main reason for this is that the standing stocks (B_i) and instantaneous flow rates (Q/B) and (P/B) are those usually estimated, they are system size independent and therefore comparable between systems, and one can relate to them. That's much more difficult for absolute values. Once inside Ecopath, it is, however, the absolute flow rates that are used in the calculations, but that's a different story.

The production equation, aka Master Equation 2 (Eq. 2) can similarly be expressed as,

$$P_i = M2_i \cdot B_i + C_i + BA_i + E_i + M0_i \cdot B_i \tag{4}$$

where $M2_i$ is the predation mortality (year^{-1}), and $M0_i$ is an "other mortality" instantaneous rate (year^{-1}), both of which becomes flow rates (t km^{-2} year^{-1}) when multiplied with biomass (t km^{-2}). The parameters are again explained in Tables 1 and 2.

"Other mortality" is often called $M0$ in some models – dating back to the MSVPA (and probably Andersen and Ursin's North Sea model), and we have adopted this convention.

Other mortality includes mortality due to diseases, starvation, etc.. The animals or plants concerned will become flow to detritus. In addition, mortality caused by predator groups not explicitly included in the model are included in the $M0$ term. This mortality term is in the MSVPA called $M1$, while in EwE it is included in $M0$ as Ecopath models traditionally would be descriptive and inclusive (and hence $M1$ is likely to be small).

For MICE type models, one should be aware that the M1-part of the M0 flow doesn't actually go to detritus, but is being consumed by predators not included in the model. Given that MICE models are focused on a specific research question this is not likely to be of concern.

The "other mortality" is the difference between total production and the sum of export, biomass accumulation, net migration, and predation mortality.

The "other mortality" thus expresses the mortality terms that the Ecopath model does not include, it could for instance be fish dying of diseases or old age, or mortality due to predators not considered in the model. It follows that 1-$M0_i$ expresses the proportion of the production for which the fate is described in the model. We call that entity the "ecotrophic efficiency" (EE_i) in tradition with Polovina's first Ecopath model[4], and it can be expressed,

$$EE_i = \frac{M2_i \cdot B_i + C_i + BA_i + E_i}{P_i} \tag{5}$$

In Eq. 4 and Eq. 5 all terms are expressed as flow rates (t km^{-2} year^{-1}). If these flow terms are made relative to biomass (t km^{-2}), and considering that $F_i = C_i/B_i$, they become rates (year^{-1}), and as $F_i = C_i/B_i$, Eq. 4 can be re-expressed as,

$$\left(\frac{P}{B}\right)_i = M2_i + F_i + \frac{BA_i}{B_i} + \frac{E_i}{B_i} + M0_i \tag{6}$$

An interesting twist to Eq. 6 is that the Ecopath mortality form (at *Ecopath > Output > Mortality*) actually shows this equation.

This equation is important, study it carefully. We describe production as the sum of predation mortality plus fishing mortality plus net migration plus biomass accumulation plus "other mortality".

Oh, that's actually the second Master Equation (Eq. 2), we're back where we started, neat.

The following table provides an overview of the input parameters for Ecopath models.

Table 1. Basic input parameters for Ecopath models

Input parameter	Name	Default value	Unit
B_i	Biomass		t km^{-2}
$(P/B)_i$	Production/biomass ratio		year^{-1}
$(Q/B)_i$	Consumption/biomass ratio		year^{-1}
EE_i	Ecotrophic efficiency ($EE_i = 1 - M0_i$)		(proportion)
BA_i	Biomass accumulation	0	t km^{-2} year^{-1}
E_i	Net migration (immigration - emigration)	0	t km^{-2} year^{-1}

Table 2. Other input parameters for Ecopath models

Input parameter	Name	Default value	Unit
DC_{ji}	Proportion of prey i in diet of predator j	0	(proportion)
U_i	Unassimilated part (excretion + egestion)	0.2	(proportion)
C_i	Catches by fleet	0	t km^{-2} year^{1}

Table 3. Estimated parameters for Ecopath models

Parameter	Name	Unit
P_i	Production ($P_i = B_i\,(P/B)_i$)	t km^{-2}year^{-1}
Q_i	Consumption ($Q_i = B_i\,(Q/B)_i$)	t km^{-2}year^{-1}
g_i	Gross food conversion efficiency $g_i = P_i / Q_i$, can be an input in which case either $(P/B)_i$ or $(Q/B)_i$ is estimated	(proportion)
R_i	Respiration ($= Q_i - P_i$ - unassimilated food)	(proportion)
F_i	Fishing mortality ($F_i = C_i / B_i$)	year^{-1}
MO_i	Other mortality ($MO_i = 1 - EE_i$)	year^{-1}
$M2_i$	Predation mortality ($M2_i = \sum B_j\,DC_{ji}$)	year^{-1}

Parameter estimation

Not all parameters used to construct a model need to be entered. The Ecopath model "links" the production of each group with the consumption of all groups, and uses the linkages to estimate missing parameters, based on the mass-balance requirement of the second Ecopath Master Equation Eq. 2 and Eq. 4, that production from any of the groups has to end somewhere else in the system. Ecopath balances the system using one production equation for each group in the system. For a system with n groups, n production equations as in Eq. 4 are used,

$$B_1\left(\frac{P}{B}\right)_1 EE_1 - B_1\left(\frac{Q}{B}\right)_1 DC_{11} - B_2\left(\frac{Q}{B}\right)_2 DC_{21} \ldots - B_n\left(\frac{Q}{B}\right)_n DC_{n1} - Y_1 - E_1 - BA_1 = 0$$
$$B_2\left(\frac{P}{B}\right)_2 EE_2 - B_1\left(\frac{Q}{B}\right)_1 DC_{12} - B_2\left(\frac{Q}{B}\right)_2 DC_{22} \ldots - B_n\left(\frac{Q}{B}\right)_n DC_{n2} - Y_2 - E_2 - BA_2 = 0$$
$$B_3\left(\frac{P}{B}\right)_3 EE_3 - B_1\left(\frac{Q}{B}\right)_1 DC_{13} - B_2\left(\frac{Q}{B}\right)_2 DC_{23} \ldots - B_n\left(\frac{Q}{B}\right)_n DC_{n3} - Y_3 - E_3 - BA_3 = 0 \quad (7)$$
$$\vdots$$
$$B_n\left(\frac{P}{B}\right)_n EE_n - B_1\left(\frac{Q}{B}\right)_1 DC_{1n} - B_2\left(\frac{Q}{B}\right)_2 DC_{2n} \ldots - B_n\left(\frac{Q}{B}\right)_n DC_{nn} - Y_n - E_n - BA_n = 0$$

where the parameters are as in Tables 1 and 2. A system of linear equation as in Eq. 7 can be solved using standard matrix algebra – you may have learned that in precalculus or algebra classes. If, however, the determinant of a matrix is zero or if the matrix is not square, it has no ordinary inverse. Still, a generalized inverse can be found in most cases. For Ecopath, we have adopted an approach described by McKay[5] to estimate the generalized inverse. If the system of linear equations is overdetermined (more equations than unknowns), and the equations are not mutually consistent, the generalized inverse method provides least square estimates to minimize discrepancies[6]. While the generalized inverse in principle is a great way of solving a system of linear equations, it is in practice not used much in the Ecopath mass-balance routine. By iteration through the system, it is usually possible to solve many of the equations. Those equations are eliminated and the inversion is only used where and if needed.

An important implication of the mass-balance equation Eq. 7 is that information about predator consumption rates and diets concerning a given prey can be used to estimate the predation mortality term for the group, or, alternatively, that if the predation mortality for a given prey is known the equation can be used to estimate the consumption rates for one or more predators instead.

The gross food conversion efficiency, g_i, is estimated using

$$g_i = \frac{(P/B)_i}{(Q/B)_i} \tag{8}$$

while Q/B are attempted solved by inverting the same equation. The P/B ratio is then estimated (if possible) from

$$\left(\frac{P}{B}\right)_i = \frac{\sum_{j=1}^{n} Q_j \cdot DC_{ji} + C_i + E_i + BA_i}{B_i \cdot EE_i} \tag{9}$$

This expression can be solved if both the catch, biomass and ecotrophic efficiency of group i, and the biomasses and consumption rates of all predators on group i are known (including group i if a zero order cycle, i.e., "cannibalism" exists). The catch, net migration and biomass accumulation rates are required input, and hence always known;

The EE is estimated from

$$EE_i = \frac{M2_i \cdot B_i + C_i + E_i + BA_i}{P_i} \tag{10}$$

where the predation mortality $M2_i$ is estimated as in Table 3 (= the first term of the numerator in Eq. 9.

In cases where all input parameters have been estimated for all prey for a given predator group it is possible to estimate both the biomass and consumption/biomass ratio for such a predator. The details of this are described in the original Ecopath II User Guide Appendix 4, Algorithm 3.

If for a group the total predation can be estimated it is possible to calculate the biomass for the group as described in detail in the original Ecopath II User Guide, Appendix 4, Algorithm 4.

In cases where for a given predator j the P/B, B, and EE are known for all prey, and where all predation on these prey apart from that caused by predator j is known the B or Q/B for the predator may be estimated directly.

In cases where for a given prey the P/B, B, EE are known and where the only unknown predation is due to one predator whose B or Q/B is unknown, it may be possible to estimate the B or Q/B of the prey in question.

Once the loop no longer results in estimate of any missing parameters a set of linear equations is set up including the groups for which parameters are still missing. The set of linear equations is then solved using a generalized inverse method for matrix inversion described by Mackay[7]. It is usually possible to estimate P/B and EE values for groups without resorting to including such groups in the set of linear equations.

The loop above serves to minimize the computations associated with establishing mass-balance in Ecopath. The desired situation is, however, that the biomasses, production/biomass and consumption/biomass ratios are entered for all groups and that only the ecotrophic efficiency is estimated, given that no procedure exists for its field estimation.

Indeed, the central point in this is that the system of linear equations in Eq. 7 can be solved for one unknown parameters for each equation. So, the advice is to leave one input parameter unknown for each group in the model, and that one parameters is preferably EE, unless no biomass estimated is available. More about that next.

Guidelines for parameter estimation

The parameters in Table 2, i.e. the diets (DC), the unassimilated part (U) and the catches (C) must always be entered as Ecopath input along with one of the six parameters in Table 1, i.e. biomass (B), production/consumption ratio (P/B), consumption/biomass ratio (Q/B), ecotrophic efficiency (EE), biomass accumulation (BA), and net migration (E). When running the Ecopath parameterization, the program will if all four basic input parameters, (B, P/B, Q/B, and EE) are entered, ask if you want to estimate biomass accumulation (BA)? If you answer no, it will ask if you want to estimate net migration (E)?

While the matrix inversion used for solving for missing parameters in Eq. 7 is flexible, it is a flexibility that should be used carefully. so a few guidelines.

Guidelines

Unless you have reason for doing it differently, leave the biomass accumulation and net migration at the default value (0).

We have a good idea of Q/B ratios for basically all kinds of organisms, so don't let the program estimate Q/B

P/B values (year^{-1}) relates to the average longevity (B/P, year) and to standard assessment outputs (Z, year^{-1}), so should not need to be estimated.

If biomass estimates are available, use them and estimate EE.

> If you don't have biomass estimates, guess a reasonable *EE* value.

Note that it is generally not possible to estimate *B* or *P/B* for apex predators from which there are no predators or catches. The tutorial about mass balance can give you some hands-on experience to get started.

Attribution

This chapter is in part adapted from the unpublished EwE User Guide: Christensen V, C Walters, D Pauly, R Forrest. Ecopath with Ecosim. User Guide. November 2008.

Notes

1. This figure was made in the early 1990s, and we haven't updated it for sentimental reasons (even though it would look much better with current technology). It tells the story to be told.

2. Winberg, G. G., 1956. Ratę of metabolism and food requirements of fishes. Nauchnye Trudy Belorusskogo Gosudarst- vennogo Universiteta. Mińsk., 253 pp. (Transl. from Russian by J. Fish. Res. Bd Can. Transl. Ser. 194, 1960). https://waves-vagues.dfo-mpo.gc.ca/library-bibliotheque/38248.pdf

3. Essington TE. 2007. Evaluating the sensitivity of a trophic mass balance model (Ecopath) to imprecise data inputs. CJFAS 64: 628-637 https://doi.org/10.1139/f07-04

4. Polovina, J.J., 1984. Model of a coral reef ecosystem. Coral Reefs 3, 1–11. https://doi.org/10.1007/BF00306135

5. Mackay A. 1981. The generalized inverse. Practical Computing, September p. 108-110

6. Christensen, V., Pauly, D., 1992. ECOPATH II — a software for balancing steady-state ecosystem models and calculating network characteristics. Ecological Modelling 61, 169–185. href="https://doi.org/10.1016/0304-3800(92)90016-8">https://doi.org/10.1016/0304-3800(92)90016-8

7. Mackay, *op. cit.*

6.

Biomasses and units

Though biomasses to some extent can relate to "real estate" the issue is not "location, location, location", but "units, units, units". When you work with ecosystem models, you'll have to obtain information for a multitude of sources and they will be using different units that need to be converted – and that often leads to conversion errors.

Units are important, and one really important thing that Ecopath has contributed has been to force (or maybe entice is a nicer word) modellers to standardize biomasses to a per unit area basis. So, for biomasses, the standard unit is ton per square kilometre in Ecopath. That makes it straightforward to compare abundance between ecosystems, where using total tons in the overall ecosystem does not just because of differences in physical ecosystem size.

How do you then get biomasses for your model? Fortunately, biomasses are standard output from surveys and assessments, and we will refer to that literature without being much more specific about how to obtain biomasses. As a rough classification, note that there are "direct" estimation methods such as trawl swept area, acoustic target expansion, visual census, plankton sampling, and "indirect" methods where biomass are output from assessments that use multiple sources of information to estimate biomasses. The biomasses add constraints to your model, and constraints make the model outputs appear to be less uncertain.

If at all possible, get biomasses from local sources (i.e. for your ecosystem), and be aware that biomasses "don't travel well". It helps that we are using per unit area biomasses, but conditions really vary from system to system due notably to differences in productivity and fishing pressure over time.

The million dollar question

How do you convert from $t\ km^{-2}$ to $g\ m^{-2}$? The answer is: they are equal.

So, when you evaluate model parameters, think $t\ km^{-2}$ for the big things, and $g\ m^{-2}$ for the small.

For instance, this bay is around $100\ km^2$ and there are some 100 seals each with a weight of 50 kg. That's $50\cdot100$ kg = 5 t in $100\ km^2$ = $0.05\ t\ km^{-2}$. Or, if we assume

---

there's 3 shrimps per m^2, each weighing 2 g; then the biomass is 6 g m^{-2} = 6 t km^{-2} as we know now.

That simple conversion between t km^{-2} and g m^{-2} really makes it simple and elegant to relate to biomass estimates for all kind of critters in an ecosystem. But watch out because biomasses for smaller critters (zooplankters, benthic invertebrates, insect larvae) are often reported in "dry weight" units without information on the drying protocol, and must be converted to the wet weight units typically used for larger critters like fish. The wet/dry weight ratio can vary from as low as 5 to over 10.

There is an ongoing controversy about whether one can or should use output from one model (assessment) as input for another model (ecosystem) where the sentiment from assessment scientists may be a *No!* and that we should instead use the same input (surveys) as used for assessments.

This argument ignores the fact that *all* biomass estimates are in fact based on models, i.e. on various aggregation, transformation, and calibration scaling operations applied to raw data. So working with the raw data as inputs would mean not just repeating one assessment but all estimations done for the ecosystem, including estimates of primary productivity and other supposedly "direct" measurements. Primary productivity (phytoplankton biomass) estimates for marine systems in particular are typically based on complex models evaluating satellite information, with calculations so specialized that it would make no sense to try to repeat them. Acoustic abundance estimates are similarly complex expansions of raw target data. Even simple swept area or volume conversions from nets are fraught with uncertainties about conversion factors. The same uncertainties more obviously hold true for assessment models, for which we have to at least initially hope that the panels reviewing the models have weeded out really bad estimates.

But thankfully, the Ecopath biomass estimates are not carved in stone; just as we typically do with single species assessment models, we can vary the Ecopath input values and examine how that variation affects dynamics, policy responses, and likelihood measures of goodness of fit to available time series data on relative abundance trends and outputs (like catches).

There are at least two good reasons to use most recent biomass estimates from single species models as the Ecopath input biomass estimates. First, those Ecopath estimates are used to initialize time simulations with Ecosim, providing a capability to predict forward from the most recent assessment estimates using a model that is initially consistent with the assessment model but explicitly represents trophic interactions evident in the Ecopath inputs when looking forward over time. Second,

Ecopath can provide a credibility check on the single-species model estimate, in particular whether the estimate is high enough to support estimated predation rates on it (Ecotrophic efficiency less than 1.0), and whether prey abundance is high enough to support its estimated food consumption. But be warned: it is not so easy to defend the use of estimated biomasses for early years from single species assessments to set Ecopath base biomasses for such early years, because of uncertainties in the single species results about cumulative net depletion of stock size over time due to historical removals and other factors.

7.

Production/biomass

Production refers to the elaboration of tissue (whether it survives or not) by a group over the period considered, expressed in whatever currency that has been selected. Total mortality, under the condition assumed for the construction of mass-balance models, equal to production over biomass (Allen, 1971[1]). Therefore, one can use estimates of total mortality (Z) as input values for the production over biomass ratio (P/B) in Ecopath models. Some examples of how to obtain P/B values are given below.

For multi-stanza groups, the production term is actually the total mortality term, Z for each stanza. So, if as an example, you have a juvenile stanza group and use a bioenergetic model to calculate the production, you should subtract the amount that is recruited to the next (older) stanza from the production in order to get the actual mortality, which is what Ecosim needs to work with.

Total mortality catch curves

Total mortalities can be estimated from catch curves, i.e., from catch composition data, either in terms of age-structured or of length-converted catch curves. The estimation can be carried out using appropriate software for analysis, but require careful consideration of how representative the samples are of the entire population, and the impact of related bias on the estimated mortality parameters.

Total mortality from sum of components

The P/B rate can be estimated as the sum of natural mortality ($M = M0 + M2$, assuming that the $M1$ term is included in $M0$) and fishing mortality (F), i.e., $Z = M + F$ ignoring here potential migration and biomass accumulation. In the absence of catch-at-age data from an unexploited population, natural mortality for finfish can be estimated from an empirical relationship (Pauly, 1980[2]) linking M, two parameters of the von Bertalanffy Growth Function (VBGF) and mean environmental temperature, i.e.,

$$M = K^{0.65} \cdot L_\infty^{-0.279} \cdot T_c^{0.462} \tag{1}$$

where, M is the natural mortality (year^{-1}), K is the curvature parameter of the VBGF (year^{-1}), L_∞ is the asymptotic length (total length, cm), and T_c is the mean habitat (water) temperature, in °C .

In equilibrium situations, fishing mortality (F, year^{-1}) can be estimated directly from the catch (C, including discards, t km^{-2} year^{-1}) and biomass (B, t km^{-2})

$$F = C/B \qquad (2)$$

Total mortality from average length

Beverton and Holt (1957[3]) showed that total mortality ($Z = P/B$, year-1), in fish population whose individuals grow according to the von Bertalanffy Growth Function (VBGF), can be expressed as

$$Z = P/B = \frac{K \cdot (L_\infty - \bar{L})}{\bar{L} - L'} \qquad (3)$$

where K is the VBGF curvature parameter (year^{-1}, expressing the rate at which L_∞ is approached), L_∞ is the asymptotic length, i.e., the mean size the individuals in the population would reach if they were to live and grow indefinitely, $\bar{L}$ is the mean length in catches, and L' represents the mean length at entry into the fishery, assuming knife-edge selection. Note that the $\bar{L}–L'$ denominator must be positive.

Total mortality and longevity

Mortality rates ($P/B = Z$) are not just nuisance parameters, they really mean something that one can relate to. How much does a population produce relative to its biomass? A lot for plankton and not very much for whales, right?

Mortality rates thus relates to size, e.g., for a small (1-2 mm) zooplankton like Acartia tonsa, the P/B is up around 45 year^{-1}. The much larger Calanus finmarchicus can live for several years and may have a P/B closer to 7 or 8 year^{-1}. Whales? P/B will be below 0.1 year^{-1}.

A good way to relate to such numbers is to turn them on their head. That is, think of the B/P ratio (year) to get a sense for the P/B (year^{-1}) ratio. So if a blue whale has a P/B of 0.025 year^{-1}, the inverse B/P is 40 years – that's the average longevity of blue whales (if P/B indeed is 0.025 year^{-1}). Seals with P/B of 0.14 year-1 would have an average longevity of 7 years, cod with a P/B of 0.25 year^{-1} an average longevity of 4 years, and anchovy with P/B of 2.0 year^{-1} would on average live half a year.

It makes sense, and longevity provides a good handle for evaluating what reasonable estimates of P/B may be.

There is a Quick guide on how to calculate P/B and Q/B for EwE models by Daniel Vilas, Marta Coll, Chiara Piroddi, Jeroen Steenbeek, developed for the EC Safenet Project, available for download.

Attribution

This chapter is in part adapted from the unpublished EwE User Guide: Christensen V, C Walters, D Pauly, R Forrest. Ecopath with Ecosim. User Guide. November 2008.

Notes

1. Allen, K. R. 1971. Relation between production and biomass. J. Fish. Res. Board Can., 28:1573-1581. doi 10.1139/f71-236

2. Pauly, D. 1980. On the interrelationships between natural mortality, growth parameters, and mean environmental temperature in 175 fish stocks. J. Cons. int. Explor. Mer, 39:175-192. https://doi.org/10.1093/icesjms/39.2.175

3. Beverton, R. J. H., and Holt, S. J., 1957. On the Dynamics of Exploited Fish Populations. Chapman and Hall, Facsimile reprint 1993, London. 533 pp.

8.

Consumption/biomass

Consumption (Q, t km^{-2} year^{-1}) is the annual intake of food by a consumer group, and it is in EwE estimated as the product of the group's biomass (B, t km^{-2}) and consumption/biomass ratio (Q/B, year^{-1}). To estimate consumption, we thus need to obtain estimates of B and Q/B for the consumer groups in models.

There are various approaches for estimating Q/B, and they can be split in (i) analytical methods and (ii) empirical methods:

(i) The analytical methods involve estimation of ration, pertaining to one or several size/age classes, and their subsequent extrapolation to a wide range of size/age classes, representing an age-structured population exposed to a constant or variable mortality. The required estimates of ration can be obtained from laboratory experiments, from studies of the dynamics of stomach contents in nature, or by combining laboratory and field data. There is an expanse of literature on this, to which we refer.

Characteristic for these methods is that they are resource- and time-consuming, and it is indeed not practical to set up laboratory or field experiment to estimate Q/B for all species or functional groups in an ecosystem model. Instead we rely on the second avenue, empirical combinations – along with estimates from analytical studies, where available.

(ii) There are a number of empirical regressions for prediction of Q/B from some easy-to-quantify characteristics of the animals for which the Q/B values are required.

Palomares and Pauly (1989[1]; 1998[2]) described based on a data set of relative food-consumption estimates (Q/B) of marine and freshwater population (n=108 populations, 38 species) a predictive model for Q/B using asymptotic weight, habitat temperature, a morphological variable and food type as independent variables. Salinity was not found to affect Q/B in fish well adapted to fresh or saltwater (other things being equal). In contrast the total mortality (Z, per year) showed a strong, positive effect on Q/B and also on the gross food-conversion efficiency (defined by $GE = Z/(Q/B)$), by affecting the ratio of small to large fish.

The authors presented three related models:

$$\log(Q/B) = 7.964 - 0.204 \log W_\infty - 1.965\, T' + 0.083\, A + 0.532\, h + 0.398\, d \qquad (1)$$

where, W_∞ is the asymptotic weight (g), T' is an expression for the mean annual temperature of the water body, expressed using $T' = 1000/\text{Kelvin}$ (Kelvin = °C + 273.15), A is the aspect ratio (see Figure 2.1), h expresses food type (1 for herbivores, and 0 for detritivores and carnivores), and d is also expressing food type (1 for detritivores, and 0 for herbivores and carnivores)

The equation can be modified to investigate the effect on mortality on Q/B, and to derive predictive models of Q/B taking explicit account of different mortalities, values of Q/B were calculated using the equation above for mortalities corresponding to $f \cdot M$, where f is a multiplicative factor with value of 0.5, 1, 2 or 4, and M is the natural mortality rate that is estimated from Pauly's (1980) empirical relationship.

$$\log(Q/B) = 8.056 + 0.300 \log f - 0.201 \log W_\infty - 1.989\, T' + 0.081\, A + 0.532\, h + 0.393\, d \quad (2)$$

where f is the multiplicative factor introduced above, and the rest of the variables are as defined earlier.

For cases where estimated of total mortality, Z, (year^{-1}) are available, the following relation may be used:

$$\log(Q/B) = 5.847 + 0.280 \log Z - 0.152 \log W_\infty - 1.360\, T' + 0.062\, A + 0.510\, h + 0.390\, d \quad (3)$$

These relationships can be used only for fish groups that use their caudal fin as the (main) organ of propulsion.

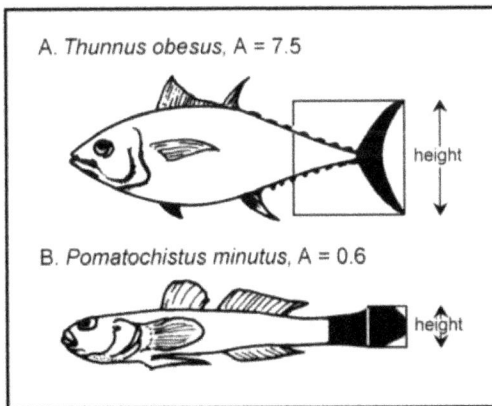

A. *Thunnus obesus*, A = 7.5

height

B. *Pomatochistus minutus*, A = 0.6

height

Figure 1 Schematic representation of method to estimate the aspect ratio ($A_r = h^2/s$) of the caudal fin of fish, given fin height (h) and surface area (s, in black).

Consumption/biomass ratios for fish are available in FishBase at the Life History tables for many species. Where analytical estimates are available those are included, while for species without such, there instead is an empirical relationship based on the equations above, see Figure 2.

Enter Winf, temperature, aspect ratio (A), and food type to estimate Q/B

Winf = 2899.2   g   Temp. = 14.0   °C

Food consumption (Q/B):   3.3   times the body weight per year

A = 1.32

Recalculate

| Detrivore | Herbivore | Omnivore | Carnivore |

Figure 2. FishBase Life History Tool for estimating Q/B from empirical relationship.

There is a *Quick guide on how to calculate P/B and Q/B for EwE models* by Daniel Vilas, Marta Coll, Chiara Piroddi, Jeroen Steenbeek, developed for the EC Safenet Project, available for download.

Attribution

This chapter is in part adapted from the unpublished EwE User Guide: Christensen V, C Walters, D Pauly, R Forrest. Ecopath with Ecosim. User Guide. November 2008.

Media Attributions

- www.fishbase.org
- www.fishbase.org Life history tool

Notes

1. Palomares, M. L. D., and Pauly, D. 1989. A multiple regression model for predicting the food consumption of marine fish populations. Aust. J. Mar. Freshwat. Res., 40:259-273. https://doi.org/10.1071/MF9890259

2. Palomares, M. L. D., and Pauly, D. 1998. Predicting food consumption of fish populations as functions of mortality, food type, morphometrics, temperature and salinity. Marine & Freshwater Research, 49(5):447- 453.https://doi.org/10.1071/MF98015

9.

Ecotrophic efficiency

Ecotrophic efficiency (EE) was defined by Ricker[1] as the proportion of a prey's production that is consumed by predators. Polovina used the term for the original Ecopath model of the unexploited French Frigate Shoals.[2]

Subsequently, we have for EwE modified the term to include exports from the system, (which notably are due to fisheries). Based on the second Ecopath Master Equation (See The energy balance of a box chapter) we have,

$$EE_i = \frac{M2_i \cdot B_i + C_i + BA_i + E_i}{P_i} \tag{1}$$

so, EE can be estimated as the ratio between the summed predation $M2$, catch C, biomass accumulation BA and net migration E relative to the production P for any group i.

If your model is descriptive (as Polovina's model was), your aim likely is to describe the energy flow in the entire ecosystem. If that's the case, the EE indeed is an "ecotrophic efficiency" that describes the proportion of the energy produced by a group that it passed on through the trophic web or exported (e.g., through fisheries). So, if EE is 0.95 then 95% of the production of the group is passed on to predators or fisheries.

But what about predictive (or MICE) models? Such models tend to be focused on the specific policy/research questions they are built to address, and as such may not give a complete picture of the food web interactions in the given ecosystem. That boils down to there being a considerable amount of unexplained mortality (M1) in the model. In such cases the EE isn't really an "ecotrophic efficiency". Hence, the following may be a better way to grasp EE.

> EE expresses the proportion of the mortality for which the model describes the fate.

Estimating EE

It is difficult to estimate EE independently, and few, if any, direct estimates appear to exist. Recognizing this, an *EE* of 0.95, based on Ricker (1968) was used for many groups in Polovina's original model[3] and in a number of later models.

Evaluating EE

Intuitively one would expect *EE* to be very close to 1 for small prey organisms, diseases and starvation probably being, for such groups, much less frequent than predation. For some groups, *EE*, may however, be low.

It is often seen that phytoplankton simply die off (as "snow") in systems where blooms occur (*EE* of 0.5 or less). Also, kelps and seagrasses are hardly consumed when alive (*EE* of 0.1 or so), and apex predators have very low *EE*s when fishing intensity is low. There are indeed many incidences of tunas or cetaceans simply dying and sinking reported from open oceans, with abyssal organisms (e.g., ratfishes) specialized in feeding on such carcasses.

If *EE*s are estimated it is often because of lack of biomasses. It should not be because of lack of *P/B* or *Q/B* values – it is better to guess those than to let them be estimated from the Ecopath mass balance. When biomasses are estimated, one needs to carefully examine how realistic those biomasses are. We've seen examples where there were biomasses entered for only unexploited high trophic level groups, and everything else being estimated from guessed *EE*s. So, if you have to used *EE*s as input, check out PreBal[4] and compare the estimated biomasses to estimates from similar ecosystems.

Notes

1. Ricker WE. 1969. Food from the Sea. pp 87-108 in: Cloud P (chair) Resources and man, a study and recommendations. Report of the Committee on Resources and Man. US Natl Acad Sci. Freeman, San Francisco, California

2. Polovina, J.J. 1984. Model of a coral reef ecosystem. Coral Reefs 3, 1–11. https://doi.org/10.1007/BF00306135

3. Polovina, J.J. (1984) *op. cit.* https://doi.org/10.1007/BF00306135

4. Link JS. 2010. Adding rigor to ecological network models by evaluating a set of pre-balance diagnostics: A plea for PREBAL, Ecological Modelling, 221(2): 1580-1591, https://doi.org/10.1016/j.ecolmodel.2010.03.012.

10.

Mass balance

Mass balance is performed using a number of algorithms and a routine for matrix inversion, see the energy balance of a box for a description of these. Once the program has estimated the missing parameters, the system balances the input and output of each group, using respiration for adjustments. The relationship used is

Master Equation 1:

$$Consumption = production + respiration + unassimilated\,part \qquad (1)$$

where, consumption is the total consumption for a group, i.e., biomass · (consumption / biomass). Respiration is the part of the consumption that is not used for production or recycled as egestion or excretion. Respiration is nonusable currency, i.e., it cannot be used by the other groups in the system. Autotrophs with $Q/B = 0$ and detritus have zero respiration. Unassimilated food is an input parameter expressing the fraction of food that is not assimilated, (i.e., is egested or excreted). For models whose currency is energy, the default is 0.20, i.e. 20% of consumption for all groups, though this is most applicable for finfish groups following Winberg[1]. The non-assimilated food is directed to the detritus.

If the model currency is a nutrient, there is no respiration. Instead, the model is balanced such that the non- assimilated food equals the difference between consumption and production.

> Some consumers are also producers, e.g., coral reefs can be a bit of both. We accommodate that by noting that production in the first Master Equation does not include primary production, i.e., it is defined as biomass · (production / biomass) · (1 – PP), where PP is the proportion of total production that can be attributed to primary production. We thus have that (1 – PP) = 0 in plants, 1 in heterotrophic consumers, and intermediate in the 0 to 1 range for e.g., corals or tridacnid clams.

An exhaustive set of guidelines for how a model should be balanced cannot be given. However, if it existed, such a set would include the following general guidelines

- Make sure to document what is done in the balancing process by entering remarks for all parameters and to extract these subsequently. A model where the balancing process is not appropriately documented is not likely to be publishable;

- Remember which data that are the more reliable and avoid changing these;

- Formulate assumptions and argumentation for changes: the ones easy to explain are likely to be the better assumptions;

- Start by looking at the estimated values. Are the EE values possible (less than 1)? Are the g (= P/Q) values physiologically realistic (0.1-0.3 for most groups, perhaps lower for top predators and higher for very small organisms, (e.g., up to 0.5 for bacteria). If not decide from where the problem is the biggest if you want to balance your model starting from the bottom (producers) or from the top down;

- Search out one group with a bigger problem and try to solve this. Are the P/B, Q/B and B values appropriate for this group? What would happen to, e.g., the g and the EE if you changed the parameters? If the problem is the consumption by predators, look at the Predation mortality form, and identify the quantitatively most important predators. Check the diet compositions and B and Q/B values for these predators;

- Continue for as long as necessary, documenting carefully what changes are made. It may be a good idea to save the data file under a new name before/after making the set of changes;

- You may get warnings that the "Respiration cannot be negative". If this happens the second master equation of Ecopath has been violated. We have:
  Consumption = production + respiration + unassimilated food
  or
  $Q = P + R + U$
  Expressing this relative to consumption we have:
  $1 = P/Q + R/Q + U/Q$
  Of these P/Q is entered as the gross food conversion efficiency (g) (or estimated from entered P and Q) and U/Q is the proportion of food that is not assimilated. If $g + U/Q$ exceeds unity, then R/Q and hence the respiration, R, has to be negative. You will need to reduce the production/consumption (g) ratio by lowering the production/biomass (P/B) ratio or increasing the consumption/biomass (Q/B) ratio, and/or reduce the proportion of unassimilated food;

- Examine the respiration/biomass (R/B) ratios for each group. Generally this ratio reflects activity level. For fish it should as a rule be in the range 1-10 $year^{-1}$, for copepods perhaps around 50-100 $year^{-1}$. Please consult physiology texts for more information. If the ratio seems high it may be necessary to change the (assumed) proportion of the food that is not assimilated on the basic input form;

- Examine the Electivity form. Do the preferences seem reasonable?

- Examine the predation mortalities at *Ecopath > Output > Mortality rates > Mortalities*, along with the predation mortality spreadsheet (*Ecopath > Output > Mortality rates > Predation mortality rates*) to identify how important the various predators are for any group. Does this show what you expect? Are the predators shown to be the most important predators in accordance with what you expect? If not, re-evaluate your model's diet compositions. The information on the mortality forms is very important!

- Noting how the energy balance of a group is formulated, it is clear that, for instance, increasing the proportion of the consumption that is not assimilated will leave less energy to respiration (production being unaffected). This will result in a lower R/B ratio and a larger flow to the detritus. The latter may be necessary to balance the model if there is only little system surplus production.

Parameter evaluation

The program estimates the missing parameters and a number of indices without further input. Your model will probably not look very convincing the first time you run it. Keep an eye open for warning messages while you make your way through the forms. In the more serious cases, the parameter estimation will be aborted, and you will have to edit your data. To improve your chances of identifying problems, you will in some cases only get a warning and the program will continue.

> **What was that warning, again?**
> You can find warnings and messages in the Status panel at the lower left of the EwE form

The following sections may help you evaluate the results of a run.

Are the *EE*'s between 0 and 1?

Ecotrophic Efficiencies (*EE*) represents the proportion of the production that is "used" in the system, or to be more precise, the proportion of the production that the model describes the "fate" of. This is a parameter that is difficult to measure empirically, but it is one we can relate to. If your model is detailed with lots of predators and fisheries impacting, e.g., small pelagics, we'd expect that the *EE* for that group should be close to 1. A top predator with only low fishing pressure should have a low *EE*, and in a system with seasonal plankton blooms the *EE* for phytoplankton should be intermediate, maybe 0.5, to give a few examples.

EE should as proportion be between 0 and 1 (inclusive). A value of zero indicates that no other group or fishery consume the given group. Conversely, a value close to 1 indicates that the group is being heavily preyed upon or grazed and/or that fishing pressure is high, thus describing almost fully what happens to the group's production.

If, in a first run, any of the *EEs* are larger than 1, something is wrong: it is not possible for more to be eaten or caught than is produced. The problem can of course be due to the equilibrium assumption not being met, e.g., when the model includes a new fishery on a previously unexploited stock – in that case include a negative biomass accumulation term. So, you should have a closer look at the input parameters.

It may be worthwhile to check the food consumption of the predators, and the production estimates of the group. Compare the food intake of the predators with the production of their prey. Most often, the diet compositions will have to be changed – often the diets are more "pointers" to, than reliable estimates of the real values.

> When checking for mass-balance, the first step is to go to the *Ecopath > Output > Mortality rates > Mortalities*. This screen shows the second Master Equation, *Production = predation \ mortality + fishing \ mortality + biomass \ accumulation+ net \ migration+ other \ mortality*. Systematically check for groups with EE problems, and especially note if the issue is due to fishing or predation. If the fishing mortality ($F =$ catch/biomass) is too high, check your catches for the fleet(s) causing this. If the predation mortality ($M2$=total predation/biomass) is too high, check *Ecopath > Output > Mortality rates > Predation mortality rates*, look across the row for each of the groups with too high *EE*, and identify the predators that are causing too high *EE*. For those groups, check their diet compositions – often, a very low diet proportion for a common predator eating a rare prey causes excessive predation mortalities for that prey.

"Cannibalism" in the sense of within-group predation often causes problems. If a group contributes 10% or more to its own diet, this alone may result in consumption being higher than the production of the group. The solution to this is to split the group into juveniles and adults, with the adults acting as predator on the juveniles. The juveniles must then have a higher production rate than the adults, as production is almost always inversely related to size.

It is advisable to make one change at the time when editing input parameters. Make that one change, note down what you did, rerun the Basic estimates routine, re-examine the run, and if necessary re-edit the data, etc. Continue with one change at a time until you get a run you consider acceptable. Make sure, through the entry of remarks in the Remarks window, to record en route what you do and why.

Ecotrophic efficiency of detritus

The ecotrophic efficiency, *EE*, of a detritus group is defined as the ratio between what flows out of that group and what flows into it. Under steady-state assumption, this ratio should be equal to 1.

Estimates of *EE* of less than 1 indicate that more is entering a detritus group than is leaving it.

Estimates of *EE* of more than 1 for a detritus group also require attention. They indicate that the primary production and/or the inputs to the lower parts of the food web are too small to support consumption from that group. It will be necessary to examine the basic inputs that define production and consumption of the lower parts of the food web closely, and to examine whether more detritus should be directed to the detritus group.

Of importance for the flow to detritus is the parameter for non-assimilated food. The default value of 0.2 often underestimates egestion, especially for herbivores and detritivores. For zooplankton eating phytoplankton a value of 0.4 results in more detritus being produced and also often leads to more reasonable respiration/biomass ratios than obtained with the default excretion rate of 0.2. Higher parameter values means that a greater flow is directed to detritus and less to respiration for a given group.

Are the efficiencies possible?

Recall that the gross food conversion efficiency, *g*, is defined as the ratio between production and consumption. In most cases, production/consumption ratios will range from 0.1 to 0.3, but exceptions may occur, (e.g., bacteria, nauplii, fish larvae and other small, fast-growing organisms). If the *g* values are unrealistic, check the input parameters, especially for groups whose production has been estimated. In such cases, carefully editing the diet composition of the predators of the problem groups will generally help.

Next are some notes about some common causes of problems during balancing.

Problem 1: Loops

In cases where *P/B* is to be estimated for groups that feed on each other (cycles) the program may first estimate a *P/B* for one group based on the consumption by the other groups. Subsequently it may estimate the *P/B* for the second group based on the consumption by the first, and then it may continue with the P/B for the first again, and so on in a loop. The result may be completely unrealistic parameter estimates.

> As a rule, don't estimate *P/B* or *Q/B*, you can provide better estimates than the mass-balance routine!

It is necessary to break such loops, e.g., by entering the *P/B* for one of the groups. If all ecotrophic efficiencies are low it indicates that the trophic transfer efficiencies are low. This may be OK for a system with high production and low abundance of organisms. It may however also indicate that the estimates of the biomasses in the system are too low.

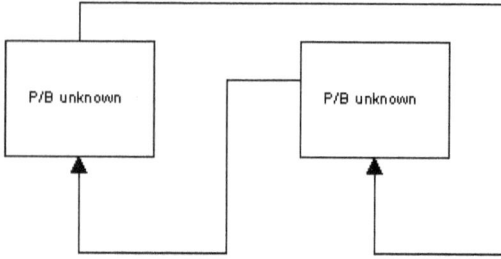

Problem 2: Cannibalism (0-order cycles)

Groups where 0-order cycles (cannibalism) are important should be broken into two or more groups. Such cases occur, for example, when a predatory fish feeds on fish of the same species or functional group. The prey fish will, however, be smaller fish, and often the P/B value for the group is based on the recruited part of the population only, and thus does not cover the dynamics of the juveniles, (which generally have much higher P/B values than the recruited part of the population). The solution may be to split the group in an adult and a juvenile fish group. This will also be an advantage for subsequent Ecosim simulations.

Remember that the gross food conversion efficiency (g) is the P/Q ratio. Typically, this ratio is in the range of 10-30%. If the proportion of the 0-order cycle is in the same range there may not be any production left over for other purposes (predation and export). As a guideline if a 0-order cycle includes more than say 5% of the diet composition it is necessary to consider if it would be better to split the group in two.

Problem 3: Estimation of predator consumption and prey production

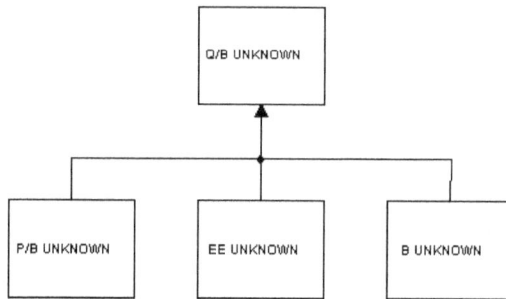

In this example it is assumed that the consumption is unknown for the predator and the (used part of the) production, (i.e., the B, P/B or EE) unknown for all of the prey groups. In this case, it will not be possible for the program to calculate meaningful parameters and it will (probably) resort to the trivial solution: set the Q/B for the predator to zero, and see what can be estimated for the other groups. The problem is easily identified from an examination of the estimated parameters and statistics. The solution may be to either input a gross efficiency for the predator or one of the missing input parameters for one of the prey groups.

Automatic mass balance?

Earlier versions of EwE (5) had an automatic mass-balance routine.[2], but we have not migrated that utility to later versions. The reason for this is GIGO = Garbage in, Garbage out. Input parameters have to checked carefully before doing mass-balance, one has to weed out error, such as notably unit conversion errors. Our experience is that that process cannot be automated. (But it was a nice and elegant routine).

Attribution: This chapter is in part adapted from the unpublished EwE User Guide: Christensen V, C Walters, D Pauly, R Forrest. Ecopath with Ecosim. User Guide. November 2008.

Notes

1. Winberg, G. G., 1956. Rate of metabolism and food requirements of fishes. In: Transl. Fish. Res. Board Can., Translation Series 194. pp. 1-253. https://waves-vagues.dfo-mpo.gc.ca/library-bibliotheque/38248.pdf

2. Kavanagh, P., N. Newlands, V. Christensen and D. Pauly. 2004. Automated parameter optimization for Ecopath ecosystem models. Ecol. Model. 172:141-149. https://doi.org/10.1016/j.ecolmodel.2003.09.004

11.

Multi-stanza life histories

Biomass-dynamics or multi-stanza?

In EwE, you can create a set of biomass groups representing life history stages or stanzas for species that have complex trophic ontogeny. Mortality rates (predation, fishing, other) and diet composition are assumed to be similar for individuals within each stanza. For instance, larvae having high mortality and feeding on zooplankton, juveniles having lower mortality and feeding on benthic insects, adults having still lower mortality and feeding on fish. To enable this feature, you must enter baseline estimates of total mortality rate Z and diet composition for each stanza, then biomass, Q/B, and BA for one "leading" stanza only.

For Ecopath mass balance calculations, the total mortality rate Z entered for each stanza-group is used to replace the Ecopath P/B for that group. That is, the second Ecopath master equation is interpreted as mass balance accounting for the mortality rate for the group ($EE \times Z$). Further, the B and Q/B for all stanza-groups besides the leading (entry) stanza are calculated before entry to Ecopath, using the assumptions that:

1. body growth for the species as a whole follows the von Bertalanffy growth curve with weight proportional to length-cubed; and

2. the species population as a whole has had relatively stable mortality and relative recruitment rate for at least a few years, and so has reached a stable age-size distribution.

Under the stable age distribution assumption, the relative number of age a animals is given by $l_a/\sum l_a$ where the sum is over all ages, and l_a is the population growth rate-corrected survivorship,

$$l_a = \exp(-\sum(Z_a - aBA/B)) \qquad (1)$$

where the sum of Z's is over all ages up to a and the BA/B term represents effect on the numbers at age of the population growth rate (e.g., the cohort born one year ago should be smaller by the $\exp(-aBA/B)$ factor than the cohort born a years ago, if the relative population growth rate has been BA/B for at least a years). Further, the relative biomass, b, of animals in stanza s should be

$$b_s = (\sum_{a\ in\ s} l_a\ w_a)/(\sum_{all\ a} l_a\ w_a) \qquad (2)$$

where $w_a = (1 - \exp(-ka^3))$ is the von Bertalanffy[1] prediction of relative body weight at age a.

Knowing the biomass, B, for one leading stanza, and the b_s for each stanza s, the biomasses for the other stanzas can be calculated by first calculating population biomass

$$B = B_{leading} / B_{leading\ s} \qquad (3)$$

then setting $B_s = b_s B$ for the other stanzas.

Q/B estimates for non-leading stanzas are calculated with a similar approach, assuming that feeding rates vary with age as the ⅔ power of body weight (a "hidden" assumption in the von Bertalanffy growth model). The internal calculations of survivorship and biomass are actually done in monthly age steps, so as to allow finer resolution than one year in the stanza biomass and mortality structure (e.g., larval and juvenile stanzas that last only one or a few months).

Here are a few implementation issues for considerations when building models with multi-stanza capability:

How many stanzas?

The main computational burden of the full representation is in Ecosim, and this burden depends on the number of age classes accounted for (calculated from K, Z for adult stanza) rather than the number of stanzas with distinct mortality/feeding patterns within the age structure. So, the best advice we can give is to err on the high side. Add stanzas for each major ontogenetic shift in habitat use and diet (though larval stages can often be ignored due to low biomass, low impact on prey, and unlikely to show density-dependent effects). If necessary additional stanzas for size-age ranges that are subject to selective fishing impacts that might cause growth overfishing under some policy scenarios (growth overfishing can be a problem whenever juvenile fish are harvested over age ranges where they display accelerating growth in body weight, so cohort biomass is still increasing over the age range being fished).

Representation of seasonality?

It is common for early juvenile stanzas to be completed within a short season each year. Yet Ecopath mass balance is based on annual average mass transfers. The initialization described above is based on "spreading" the seasonal effects evenly over the annual cycle (in monthly steps), and in practice this does not cause serious problems for the mass-balance calculation/Ecopath estimation. On entry to Ecosim, users can specify seasonal recruitment patterns and represent seasonal interaction dynamics in detail, but this often forces care in all aspects of seasonality, (e.g., in prey productivity and availability as well as juvenile abundance). Generally, we find that these more detailed calculations give about the same long term population dynamics as when recruitment is treated as seasonal, except in scenarios that involve match-mismatch variation from year to year in the timing of food availability relative to the timing of recruitment (so unless you specifically want to examine match-mismatch hypotheses, consider not bothering to include seasonality in the simulations).

Representation of stanzas that occur outside the modelled system?

It is common, especially in models for coastal ecosystems, to have species that spend only part (or none) of their time in the given system. For example, juvenile rearing may be in the modelled ecosystem, but adult foraging and harvest impacts may occur in outside areas. The preferred way to handle trophic/fishery impacts for such species in EwE is to treat part (or all) of the diet for outside-migrant stanzas as imported, rather than to model the movement into and out of the system as immigration/emigration rates. With the diet import convention, EwE will still handle overall fishery impacts at the population scale whether or not these impacts occur within the modelled system; all that will be "lost" is dynamic change in food availability (and feeding rates) and predation mortality of organisms during times when they are outside the modelled system (outside world treated as having constant trophic conditions). Often, the stanzas that reside outside the modelled system may be older fish, for which the assumption of constant resource availability and natural mortality risk may be quite reasonable. When it appears that using the diet-import convention is inappropriate due to changing trophic conditions outside the modelled system, then the modelled system should be extended to include the "outside" trophic interactions of concern.

The multi-stanza representation is quite flexible, and you may find other ways to use it for effectively representing "problem processes" in ecological systems.

Attribution This chapter is in part adapted from the unpublished EwE User Guide: Christensen V, C Walters, D Pauly, R Forrest. Ecopath with Ecosim. User Guide. November 2008.

Notes

1. Von Bertalanffy, L. (1938) A quantitative theory of organic growth (inquiries on growth laws II). Human Biology, 10, 181- 213. https://www.jstor.org/stable/41447359

12.

Uncertainty

EwE has extensive ways of treating uncertainty – but it is important to first discuss types of uncertainty.

- Background ecological variation or "noise"
    - This is variation in ecological processes that is apparently random and not strongly autocorrelated over time. In single-species dynamics and assessment, recruitment in particular typically has such variation due to unmodelled environmental factors that influence survival rate; so we may seek to predict recruitment relationships in EwE, but there will almost always be environmentally-driven variation that EwE (or any other model) does not capture.

- Parameter uncertainty
    - This is the focus for treating uncertainty in EwE – and by extension the focus of this chapter.

- Structural uncertainty
    - This can for instance be model bias: how well does the model describe the system? Typically, this can be addressed by using alternative models to make predictions about the specific management/ policy question(s) that are being addressed. Within the EwE framework, this can be done by using a suite of model formulations, e.g., spanning from MICE-type models to complex formulations with a large number of functional groups and forcing functions. Structural uncertainty can come both from how interactions are represented in the system and from "external" (the larger world within which the model is embedded) forcing factors that may or may not be recognized during model development. See also the discussion about alternative models in the research question chapter.

- Observation error
    - This, e.g., includes errors in sampling surveys that result in uncertainty in biomass estimates. Very few ecosystem models incorporate raw data such as for instance from trawl surveys. Most are processed through other models, e.g., single-species assessments, or converting satellite images to chlorophyll and primary productivity – as discussed in the Biomasses and units chapter. For this reason,

very few EwE models deal directly with observation errors – though the effects of this uncertainty become evident in measures of uncertainty about parameter values derived by examining how much the parameter values can be changed without degrading the fit to the data (likelihood functions measure this).

- Implementation error

  - In Management Strategy Evaluation (MSE) this includes errors due for example to variation in catchability coefficients and uncontrolled variation in fishing effort. See the MSE chapter for details.

Most EwE models have focused on parameter uncertainty – but how does one assign uncertainty ranges to the many parameters in an ecosystem model? Consider a typical model with 30 functional groups and 5 fisheries exploiting 5 groups each. Such a model will have over 200 input parameters in the Ecopath base model alone. That is too many to estimate individual parameter uncertainties for, and if done it would not be transparent. Assigning uncertainty based on parameter type, e.g., ±20% for biomasses, is not a realistic option. Instead, we have developed an alternative approach based on parameter "pedigree". Pedigree can be defined as "a register recording a line of ancestors". It is something that for instance a horse or dog may have, and it means we know where it came from. We use it to describe how well rooted a model is in local data, and with it, how uncertain the data are. The assumption is that local data are more reliable than regional or global data, guessed data are even less reliable, but the most uncertain are estimates that are derived from another model – notably those estimated by Ecopath mass-balance.

The EwE pedigree can be defined for the key Ecopath input parameters, that is biomass, production/biomass, consumption/biomass, diets and catches, see Figure 1. Each parameter has a set of classifications following the logic described above (the range from local data to model estimates), and a default parameter uncertainty is associated with each parameter-classification type. The default parameter uncertainties can be overwritten, but given that they are reasonable and that one will have to explain why one has changed them, most models for which users have defined pedigree for input parameters have used the default values.

There are two neat aspects to defining pedigree. By assigned pedigree index [0,1] for each parameter-classification type, we can estimate an overall pedigree [0,1] for a model, which describes how well rooted the model is in local data, and which can be compared to other ecosystem models. See Morisette (2007)[1] for a neat analysis of how the quality of input data relates to complexity and stability in ecosystems. The second aspect is that the pedigree table (see Figure 1) from the colour scale gradient directly gives an overview of how well rooted the data that are used for a model are in local data. This is a great tool for communicating what data is available from a given ecosystem. It also makes it quite clear when models feed models, i.e. when a model relies heavily on data from other models. That is generally to be avoided when possible.

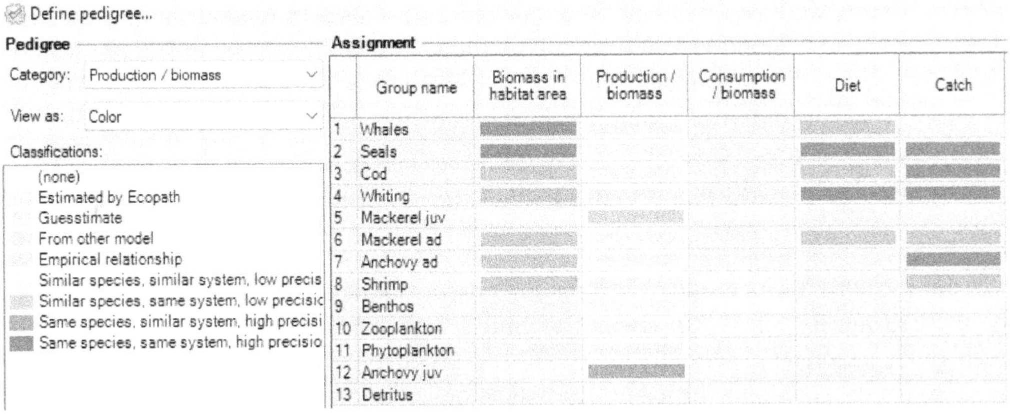

Figure 1. Input screen for defining pedigree (*Ecopath > Input > Tools > Pedigree*) for Anchovy Bay. Colours (gradient in grey scale) in the assignment table indicates classifications, such as shown in the Classification table, which is for the production/ biomass ratio. There are similar classification tables for the other input parameters, all with defined default uncertainty associated.

Earlier versions of EwE (5) had an *EcoRanger* routine, which varied each of the basic Ecopath input parameters and evaluated what impact such changes had on output parameters. We have not ported that routine to later version (6+) for the simple reason that the output wasn't interesting or credible. *EcoRanger* produced pages and pages of tables, but didn't answer any questions. Questions about uncertainty have to be related to research and policy questions (see Your research question? chapter) for uncertainty analysis to be interesting and worthwhile. Beth Fulton has illustrated this by estimating that to evaluate parameter uncertainty for all input parameters for an Atlantis model[2] would take the biggest super-computer on Earth longer than the age of the Universe. And it wouldn't be interesting! Uncertainty has to be focused on the research/policy questions that the given model is built to address.

With this in mind, it should be clear that there are no direct uncertainty analyses in the Ecopath section of EwE. That is not because there is no uncertainty associated with Ecopath parameters (as discussed above), but because Ecopath serves as a base model for other analyses, especially Ecosim. The intention with the Ecopath model is to provide one possible parameter realization for a given ecosystem. Once we have such a one, we can throw uncertainty at it by sampling, including by generating not just one, but many Ecopath and Ecosim models, thereby providing the foundation for addressing questions about uncertainty.

Sampling of parameter values can be based on Monte Carlo (MC) techniques, which is used to estimate properties of a distribution from random sampling. As an example, a MC approach can be used to draw a large number of random samples from an unknown distribution, and calculate the sample mean of those. A major benefit of MC is thus that calculating parameters for a random sample is often much easier than calculating the parameters directly from the distribution (if at all known, that is). Expanding on this,

Markov Chain MC (MCMC) is a Bayesian approach that uses random samples to generate new random samples (hence "chain") within a probability space. Each new sample depends only on the previous sample, with an incremental change (that's the "Markov property"), and the more such samples, the more the final samples will resemble the original (perhaps unknown) distribution. We refer (without specifics) to the statistics literature for details about MC and MCMC, both of which procedures are used extensively in EwE.

Ecosim includes a MC routine (*Ecosim > Tools > Monte Carlo simulation*), which can be used to run Ecosim in an attempt to find a model that improves fit to time series with given parameter uncertainty for biomasses, production/biomass ratios, consumption/ biomass ratios, ecotrophic efficiencies, diets, biomass accumulation or biomass accumulation rates, landings and discards. Parameter uncertainty can be defined for each group-parameter combination, and is often populated from pedigree. The routine will draw an Ecopath input parameter combination, evaluate the derived model based on mass-balance criteria, and if the model passes these criteria, the MC routine will run Ecosim and evaluate the fit to time series data. If a model produces a better fit to the time series data than previously obtained, the model will be retained. At the end of the run, the best fit model parameter combination can then be used as an Ecopath base model, if so desired. The output from the model runs can also be used to evaluate the overall trajectories for Ecosim runs with the given parameter uncertainty.

Where the MC routine as described above evaluates one possible Ecopath model realization at the time, the CEFAS MSE plug-in for EwE (see Management strategy evaluation chapter), provides a method for developing a suite of possible Ecopath models, which subsequently can be used for resampling in comparison of policy options.

The possibility of creating multiple Ecopath base models has also been included in the EwE Ecosampler routine, which is a very versatile approach for addressing uncertainty for any routine in Ecosim and Ecospace. Ecosampler records alternative mass-balanced Ecopath models from MC, and replays these "samples" through EwE main modules and plug-ins, including Ecopath, Ecosim, Ecotracer, EcoIndicators, value chain and others). The routine captures output variation due to base input parameter uncertainty. We here refer to the Ecosampler chapter in the EwE User Guide for details.

Uncertainty in time series (e.g., environmental forcing functions, fishing effort, or biomass series) can be addressed with the Multi-sim plug-in described in the EwE User Guide, to which we refer to details. There is a tutorial in (the web and online pdf versions of) this book, that can used to get experience with the approach (see the Uncertainty in time series data tutorial).

When it comes to Ecospace, there are only few routines and examples of explicitly evaluating uncertainty. One notable example, however, was for the Roberts Bank Terminal 2 expansion project (see the Environmental impact assessment chapter of the online version of this textbook for details) where the research question related to uncertainty was how model structure and parameterization impact predictions about ecological impact of a proposed container terminal. For this analysis we ran a complex ecosystem model 5,000 times several times over as part of very comprehensive analyses of model prediction uncertainty. The Ecospace MC approach was implemented as a plug-in and will be made available for a coming release of EwE.

As a closing point, note that EwE users need to beware of what we call the "bad apple" problem, i.e. the idea that just one bad apple can cause a whole barrel to go rotten. It doesn't always happen, it's more of an exception, but it <u>can</u> happen. That is, just because most parameters are well estimated does not mean that just one or two bad ones cannot cause the whole model to make hugely incorrect predictions – especially if used to address questions that the original model wasn't designed to answer. For example, an early model of the Georgia Strait ecosystem fit historical data very well, but had large overestimates of Q/B and P/B for Pacific hake. This went unnoticed until a later model developer "borrowed" those two parameter values and ran a scenario where harbour seal populations were reduced through harvesting. That policy test resulted (because of the bad hake parameters) in a substantial hake increase, which in turn caused the simulated herring (and two salmon) populations to be driven to extinction. To make matters worse, yet another scientist then used the bad model predictions as evidence that reducing seal populations would have counter-productive results because of the value of the seals for controlling impacts of other species like hake. The lessons from this example include (1) test your models not just to fit data, but for their specific policy predictions, (2) use pedigrees to check for possible bad parameter estimates, and (3) do not assume that effects of multiple parameters are additive so that a few errors will not have major impact on model performance.

Notes

1. Morissette L. 2007. Complexity, cost and quality of ecosystem models and their impact on resilience: a comparative analysis, with emphasis on marine mammals and the Gulf of St. Lawrence. PhD dissertation, University of British Columbia. Available at https://open.library.ubc.ca/soa/cIRcle/collections/ubctheses/831/items/1.0074903?o=12

2. See https://research.csiro.au/atlantis/

III

Network Analysis

13.

Network analysis

EwE links concepts developed by theoretical ecologists, especially the network analysis theory of Ulanowicz[1], with those used by biologists involved with fisheries, aquaculture and farming systems research. The Network analysis component of Ecopath is included as a plugin *(Ecopath > Output > Tools > Network analysis)*.

The output forms included in the plug-in include: Trophic level decomposition, Flows and biomasses, Primary production required, Mixed trophic impact, Ascendancy, Flow from detritus, Cycles and pathways, Network analysis indices in Ecosim.

Trophic level decomposition

In addition to the routine for calculation of fractional trophic levels, a routine is included in Ecopath which aggregates the entire system into discrete trophic levels sensu Lindeman. This routine, based on an approach suggested by Ulanowicz[2] (1995), reverses the routine for calculation of fractional trophic levels. Thus, for the example when a group obtains 40% of its food as a herbivore and 60% as a first-order carnivore, the corresponding fractions of the flow through the group are attributed to the herbivore level and the first consumer level.

The results of this analysis are presented in the Relative flows table under the Trophic level decomposition node (these are proportions adding up to 1). These proportions are converted to absolute amounts, presented in the Absolute flows table (t km^{-2} year^{-1} or grams of carbon m^{-2} year^{-1}), thus enabling the flows to be aggregated by trophic level and summarized in different ways.

Flows from detritus to the different model groups are calculated when you select the Flow from detritus menu item.

Transfer efficiency

Based on the trophic aggregation tables, the transfer efficiencies between successive discrete trophic levels can be calculated as the ratio between the sum of the exports from a given trophic level, plus the flow that is transferred from trophic level to the next, and the throughput on the trophic level. This is presented in a table with transfer efficiencies (%) by trophic levels.

Flows and biomasses

The absolute flows calculated in the Trophic level decomposition and Flow from detritus analyses can be aggregated to produce useful summaries by trophic level.

Primary production required

For terrestrial systems, it has been shown by Vitousek et al.[3] based on a detailed analysis of agriculture, industry and other activities, that nearly 40% of potential net primary production is used directly or indirectly by these activities. Comparable estimates for aquatic systems were not available until recently, though a rough estimate, of 2% was presented in the same publication. This figure, much lower than that for terrestrial systems, was based on the assumptions that an "average fish" feeds two trophic levels above the primary producers, and has been since revised upward.[4] The crudeness of Vitousek et al.'s approach for the aquatic systems was due mainly to lack of information on marine food webs, especially on the trophic positions of the various organisms harvested by humans. Models of trophic interactions may however help overcome this situation, and an alternative approach, based on network analysis, may be suggested for quantification of the primary productivity required to sustain harvest by humans (or by analogy by any other group that extracts production from an ecosystem).

To estimate the primary production required (PPR)[5] to sustain the catches and the consumption by the trophic groups in an ecosystem, the following procedure has been implemented in Ecopath: First, all cycles are removed from the diet compositions, and all paths in the flow network are identified using the method suggested by Ulanowicz.[6] For each path, the flows are then raised to primary production equivalents using the product of the catch, the consumption/production ratio of each path element times the proportion the next element of the path contributes to the diet of the given path element. For a simple path from trophic level (TL) I (primary producers and detritus), over TL II and III, and on to the fishery,

$$TL_I \xrightarrow{Q_{II}} TL_{II} \xrightarrow{Q_{III}} TL_{III} \xrightarrow{Q_{IV}} \text{Fishery} \tag{1}$$

the primary production (or detritus) equivalents, PPR, corresponding to the catch of Y is:

$$PPR_C = Y \cdot \frac{Q_{III}}{Y} \cdot \frac{Q_{II}}{Q_{III}} = Q_{II} \tag{2}$$

For the general (and more realistic) case where the pathways include branching the PPR corresponding to a catch Y of a given group can be quantified by summing over all pathways leading to the given group the PPR's

$$PPR_C = \sum_{\text{Paths}} \left(Y \cdot \prod_{\text{Pred,prey}} \frac{Q_{\text{Pred}}}{P_{\text{Pred}}} \cdot DC'_{\text{Pred,prey}} \right) \tag{3}$$

where P is production, Q consumption, and DC' is the diet composition for each predator/prey constellation in each path (with cycles removed from the diet compositions).

Further, the *PPR* for sustaining the consumption of each trophic group in a system can be estimated from the same equation as above by substituting the catch, Y, with the production term, P, calculated as the production/biomass ration, P/B, times the biomass, B.

PPR should actually be interpreted as flow from *TL* I as it includes primary production as well as detritus uptake. The denominator, *PP*, thus actually includes all "new" flow to the detritus groups, i.e. flow from primary producers and import of detritus.

The *PPR* is closely related to the "emergy" concept of H. T. Odum,[7] and is proportional to the "ecological footprint" of Wackernagel and Rees.[8]

Mixed trophic impact

Leontief [9] developed a method to assess the direct and indirect interactions in the economy of the USA, using what has since been called the "Leontief matrix". This approach was introduced to ecology by Hannon[10] and Hannon and Joiris.[11] Using this, it becomes possible to assess the effect that changes the biomass of a group will have on the biomass of the other groups in a system. Ulanowicz and Puccia[12] developed a similar approach, and a routine based on their method has been implemented in the Ecopath system. The "mixed trophic impact" (MTI) for living groups is calculated by constructing an n x n matrix, where the i,j^{th} element representing the interaction between the impacting group i and the impacted group j is

$$MTI_{i,j} = DC_{i,j} - FC_{ji} \qquad (4)$$

where $DC_{i,j}$ is the diet composition term expressing how much j contributes to the diet of i, and $FC_{j,i}$ is a host composition term giving the proportion of the predation on j that is due to i as a predator. When calculating the host compositions the fishing fleets are included as 'predators'.

For detritus groups the $DC_{i,j}$ terms are set to 0. For each fishing fleet a "diet composition" is calculated representing how much each group contributes to the catches, while the host composition term as mentioned above includes both predation and catches.

The diagonal elements of the *MTI* are further increased by 1, i.e.,

$$MTI_{i,i} = MTI_{i,i} + 1 \qquad (5)$$

The matrix is inversed using a standard matrix inversion routine.

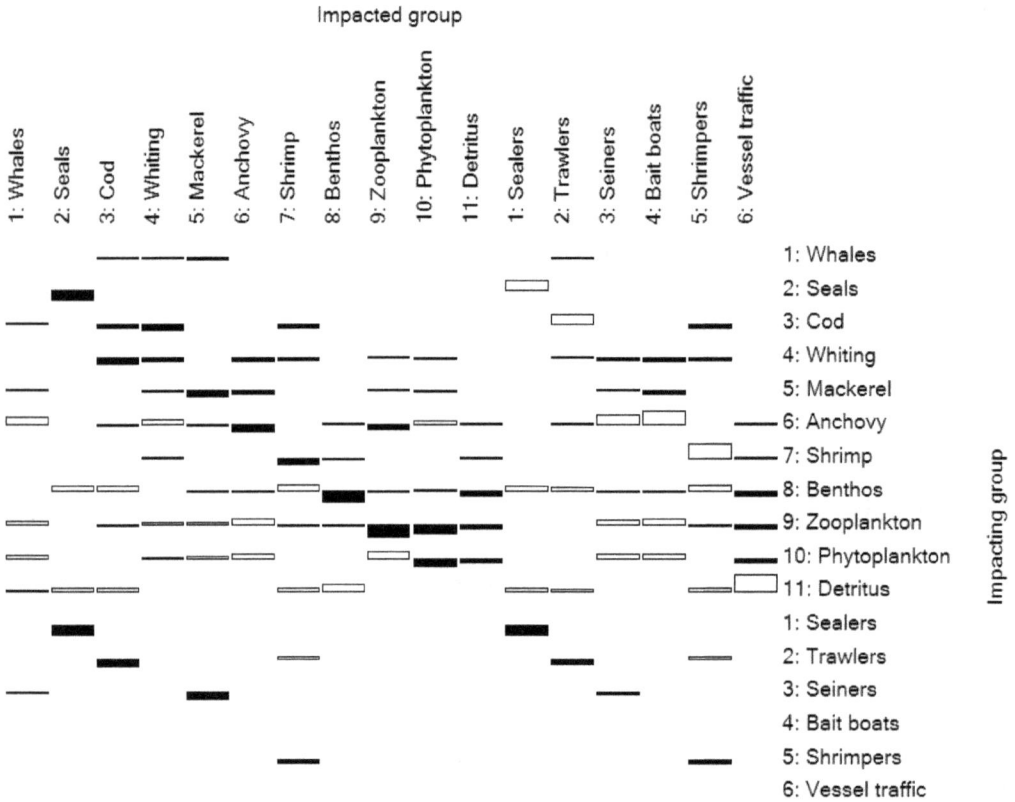

Figure 1. Mixed trophic impacts for Anchovy Bay showing the combined direct and indirect trophic impacts that an infinitesimal increase of any of the groups in the rows (to the right) is predicted to have on the groups in the columns (on top). The open bars pointing upwards indicate positive impacts, while the filled bars pointing downwards show negative impacts. The bars should not be interpreted in an absolute sense: the impacts are relative, but comparable between groups.

Note in Figure 1 that most groups have a negative impact on themselves, interpreted here as reflecting increased within-group competition for resources. Exceptions exist; thus, if a group cannibalizes itself (0-order cycle), the impact of a group on itself may be positive. In figure 1 the impact of whales on whales is negligible indicating that whales very far from their carrying capacity in the model base year.

The mixed trophic impact routine can also be regarded as a form of "ordinary" sensitivity analysis.[13] In this system, it can be concluded, e.g., that the impact of the bathypelagics on any other group is negligible: these fishes are too scarce to have any quantitative impacts. This can be seen to indicate that one need not allocate much effort in refining one's parameter estimates for this group; it may be better to concentrate on other groups.

One should regard the mixed trophic impact routine as a tool for indicating the possible impact of direct and indirect interactions (including competition) in a steady-state system, not as an instrument for making predictions of what will happen in the future if certain interaction terms are changed. The major reason for this is that changes in abundance may lead to changes in diet compositions, and this cannot be accommodated with the mixed trophic impact analysis.

Ascendancy

"Ascendency" is a measure of the average mutual information in a system, scaled by system throughput, and is derived from information theory.[14] If one knows the location of a unit of energy the uncertainty about where it will next flow to is reduced by an amount known as the average mutual information',

$$I = \sum_{i=1,j=1}^{n} f_{ij} Q_i \log(f_{ij} / \sum_{k=1}^{n} f_{kj} Q_k) \tag{6}$$

where, if T_{ij} is a measure of the energy flow from j to i, f_{ij} is the fraction of the total flow from j that is represented by T_{ij}, or,

$$f_{ij} = T_{ij} / \sum_{k=1}^{n} T_{kj} \tag{7}$$

Qi is the probability that a unit of energy passes through i, or

$$Q_i = \sum_{k=1}^{n} / \sum_{l=1,m=1}^{n} T_{lm} \tag{8}$$

Q_i is a probability and is scaled by multiplication with the total throughput of the system, T, where

$$T = \sum_{i=1,j=1}^{n} T_{i,j} \tag{9}$$

Further

$$A = T \cdot I \tag{10}$$

where A is called "ascendency". The ascendency is symmetrical and will have the same value whether calculated from input or output.

There is an upper limit for the size of the ascendency. This upper limit is called the "development capacity" and is estimated from

$$C = H \cdot T \tag{11}$$

where H is called the 'statistical entropy', and is estimated from

$$H = \sum_{i=1}^{n} Q_i \log Q_i \tag{12}$$

The difference between the capacity and the ascendency is called "system overhead". The overheads provide limits on how much the ascendency can increase and reflect the system's "strength in reserve" from which it can draw to meet unexpected perturbations.[15] As an example, the part of the ascendency that is due to imports, A0, can increase at the expense of the overheads due to imports, Q_0. This can be done by either diminishing the imports or by importing from a few major sources only. The first solution would imply that the system should starve, while the latter would render the system more dependent on a few sources of imports. The system thus does not benefit from reducing Q_0 below a certain system-specific critical level (Ulanowicz and Norden, 1990).

The ascendency, overheads and capacity can all be split into contributions from imports, internal flow, exports and dissipation (respiration). These contributions are additive.

The unit for these measures is "flowbits", or the product of flow (e.g., $t \, km^{-2} \, year^{-1}$) and bits. Here the "bit" is an information unit, corresponding to the amount of uncertainty associated with a single binary decision.

The overheads on imports and internal flows (redundancy) may be seen as a measure of system stability *sensu* Odum, and the ascendency / system throughput ratio as a measure of information, as included in Odum's attributes of ecosystem maturity. For a study of ecosystem maturity using Ecopath see Christensen 1995.[16]

Flow from detritus

The Trophic level decomposition analysis calculated the fractions of the flow from each trophic level through each model group. The Flow from detritus analysis is equivalent, but calculates the flow from detritus through each group and converts it to absolute flows ($t \, km^{-2} \, year^{-1}$).

Cycles and pathways

A routine based on an approach suggested by Ulanowicz[17] has been implemented to describe the numerous cycles and pathways that are implied by the food web representing an ecosystem.[18]

Each routine below has two forms: Pathway and Summary of pathways. The summary routine counts the number of all pathways leading from the prey to the selected consumer. The mean path length will be calculated and displayed on the form. It is calculated as the total number of trophic links divided by the number of pathways.

Consumer <- TL1

This routine lists all pathways leading from all groups on Trophic Level I (primary producers and detritus) to any selected consumer. A list of all consumers in the system will be displayed, and one can select from this. The program then searches through the diet compositions, finds all the pathways from the primary producers to the specified consumer, and then presents these pathways. Further, a summary presents the total number of pathways and the mean length of the pathways (under the Summary of pathways menu item). The latter is calculated as the total number of trophic links divided by the number of pathways.

Consumer <- prey <- TL1

This routine lists all pathways leading from all groups on Trophic Level I (primary producers and detritus) to any selected consumer via a selected prey. A pull-down list of all consumers in the system will be displayed after the heading "Pathways leading to:". Select the consumer of interest from this list then choose a specific prey from the right-hand pull-down list. The program searches through the diet compositions, finds all the pathways from the primary producers, via the selected prey, to the specified consumer, and then presents the pathways. A summary presents the total number of pathways and the mean length of the pathways (under the Summary of pathways menu item).

Top predator <- prey

Here, one enters a prey group, and the program will find all pathways leading from this prey to all top predators. A summary presents the total number of pathways and the mean length of the pathways (under the Summary of pathways menu item).

Cycles (living)

The routine identifies all cycles in the system excluding detritus and displays these, in ascending order, starting with "zero order" cycles ("cannibalism"). In addition, the total number and the mean length of the cycles will be displayed.

Cycles (all)

The routine identifies all cycles in the system and displays these, in ascending order, starting with "zero order" cycles ("cannibalism"). In addition, the total number and the mean length of the cycles will be displayed.

Cycling and path length

The "cycling index" is the fraction of an ecosystem's throughput that is recycled. This index, developed by Finn[19] (1976), is expressed here as a percentage, and quantifies one of Odum's[20] 24 properties of system maturity.[21] Recent work shows this index to strongly correlate with system maturity, resilience and stability.

In addition to Finn's cycling index, Ecopath includes a slightly modified "predatory cycling index", computed after cycles involving detritus groups have been removed.

The path length is defined as the average number of groups that an inflow or outflow passes through.[22]. It is calculated as

Path length = Total System Throughput / ($\sum Export + \sum Respiration$)

As diversity of flows and recycling is expected to increase with maturity, so is the path length. The effects of changes in the ecosystem on the network analysis indices (such as total systems throughput, Finn and predatory cycling indices, ascendency, overhead and their breakdown into various components) can then be plotted over time and compared for various scenarios of Ecosim.

Attribution: This chapter is in part adapted from the unpublished EwE User Guide: Christensen V, C Walters, D Pauly, R Forrest. Ecopath with Ecosim. User Guide. November 2008.

Notes

1. Ulanowicz, R. E., 1986. Growth and Development: Ecosystem Phenomenology. Springer Verlag (reprinted by iUniverse, 2000), New York. 203 pp.

2. Ulanowicz, R. E., 1995. Ecosystem Trophic Foundations: Lindeman Exonerata. In: Chapter 21 p. 549-560 In: B.C. Patten and S.E. Jørgensen (eds.) Complex ecology: the part-whole relation in ecosystems, Englewood Cliffs, Prentice Hall.

3. Vitousek, P. M., Ehrlich, P. R., and Ehrlich, A. H. 1986. Human appropriation of the products of photosynthesis. Bioscience, 36:368-373.

4. Pauly, D., and Christensen, V. 1995. Primary production required to sustain global fisheries. Nature, 374(6519):255-257 [Erratum in Nature, 376: 279].

5. Christensen, V., and Pauly, D., 1993. Flow characteristics of aquatic ecosystems. In: Trophic Models of Aquatic Ecosystems. pp. 338-352, Ed. by V. Christensen and D. Pauly, ICLARM Conference Proceedings 26, Manila

6. Ulanowicz, R. E., 1995. Ecosystem Trophic Foundations: Lindeman Exonerata. In: Chapter 21 p. 549-560 In: B.C. Patten and S.E. Jørgensen (eds.) Complex ecology: the part-whole relation in ecosystems, Englewood Cliffs, Prentice Hall.

7. Odum, H. T. 1988. Self-organization, transformity and information. Science, 242:1132-1139.

8. Wackernagel, M., and Rees, W., 1996. Our ecological footprint: reducing the human impact on the Earth. In: New Society Publishers. Gabriela Island. 160 p.

9. Leontief, W. W., 1951. The Structure of the U.S. Economy. Oxford University Press, New York.

10. Hannon, B. 1973. The structure of ecosystems. J. Theor. Biol., 41:535-546.

11. Hannon, B., and Joiris, C. 1989. A seasonal analysis of the southern North Sea ecosystem. Ecology, 70(6):1916-1934.

12. Ulanowicz, R. E., and Puccia, C. J. 1990. Mixed trophic impacts in ecosystems. Coenoses, 5:7-16.

13. Majkowski, J., 1982. Usefulness and applicability of sensitivity analysis in a multispecies approach to fisheries management. In: Theory and management of tropical fisheries. ICLARM Conf. Proc. 9. pp. 149- 165, Ed. by D. Pauly and G. I. Murphy

14. see Ulanowicz, R. E., and Norden, J. S. 1990. Symmetrical overhead in flow and networks. Int. J. Systems Sci., 21(2):429-437

15. Ulanowicz, 1986. *op. cit.*

16. Christensen, V. 1995. Ecosystem maturity - towards quantification. Ecological Modelling, 77(1):3-32.

17. Ulanowicz. 1986. *op.cit.*

18. For a further description see Ulanowicz, 1986, his examples 4.4 and 4.5, page 65f.)

19. Finn, J. T. 1976. Measures of ecosystem structure and function derived from analysis of flows. J. Theor. Biol., 56:363-380.

20. Odum, 1969. *op. cit.*

21. Christensen 1995. *op. cit.*

22. Finn JT. 1980. Flow analysis of models of the Hubbard Brook ecosystem. Ecology 6: 562-571.

IV

An Introduction to Ecosim

14.

Foraging arena theory

Robert NM Ahrens; Villy Christensen; and Carl J. Walters

Foraging arena theory is the driving machinery in EwE, and it represents a development that has had profound implications for making ecosystem models behave, be able to replicate the ecosystem history and make plausible predictions. Without the foraging arena theory there would be no EwE.

> There's a story about the birth of the foraging arena theory in the On modelling chapter of this book.

The foraging arena theory emerged through a series of studies during the 1990s[1] [2] [3]. The general predictions of foraging arena theory have been fairly widely used by fisheries scientists, mainly through the application of EwE-Ecosim, to explain and model responses of harvested ecosystems[4]. The potential for the underlying ecological theory upon which foraging arena theory is based to help to understand a broad range of aquatic ecosystem behaviours has apparently not been widely recognized.

Here we describe the basic models of foraging arena theory. We review the various mechanisms that can lead to these models, list the main predictions they imply, and give an overview of the practical difficulties that have been encountered in estimating critical vulnerability exchange rate parameters that appear to limit trophic interaction rates. The present chapter is an adapted extract from Ahrens et al.[5] to which we refer for a fuller presentation and notably examples with references.

Basic models of foraging arena theory

The basic assertion of foraging arena theory is that spatial and temporal restrictions in predator and prey activity cause partitioning of each prey population into vulnerable and invulnerable population components, such that predation rates are dependent on (and limited by) exchange rates between these prey components. Trophic interactions take place in the restricted "foraging arenas" where vulnerable prey can be found (Figure 1 and 2).

That is, if the total prey population is B_i, and V_i of these are vulnerable to predation at any moment (i.e. are in the foraging arena for interaction with some predator whose abundance is B_j), total prey consumption rate Q_j should be predictable as the mass action product

$$Q_j = a_{ij} V B_j \qquad (1)$$

where the predator rate of effective search a_{ij} has units of area or volume per time searched by the predator divided by the area or volume (A) of the foraging arena. Note here that Q_j is predictable as $Q_j = a_{ij} B_i B_j$ only when $V_i = B_i$, i.e. when all B_i prey and B_j predators are randomly distributed or well-mixed.

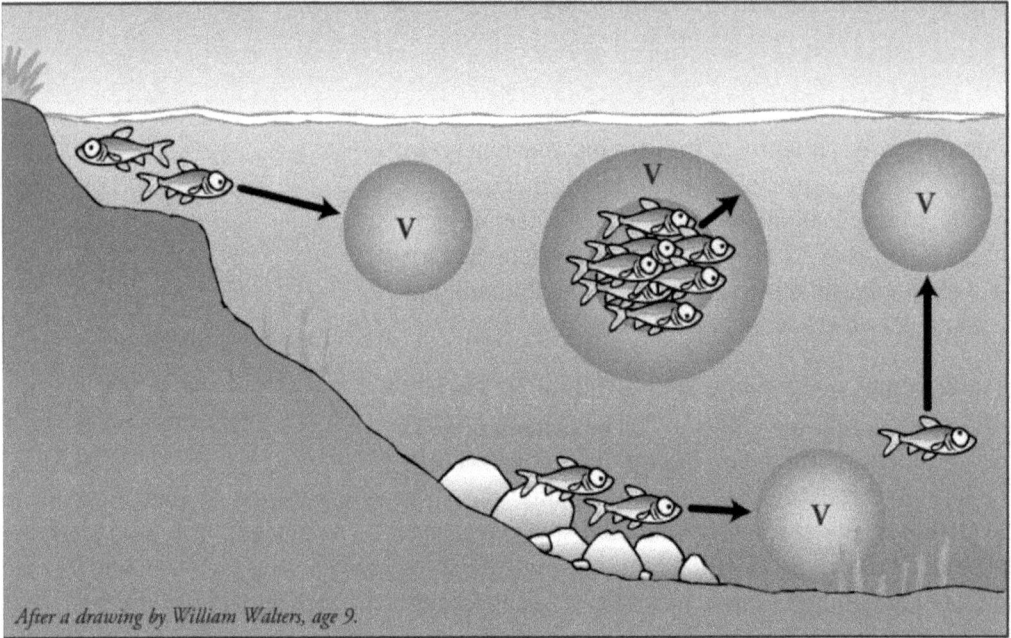

After a drawing by William Walters, age 9.

Figure 1. Aquatic organisms have evolved a diversity of behaviours that limit their exposure to predation risk. The use of spatial refuges from predation is likely to restrict foraging to limited volumes (V) nearby and limit predator-prey interaction.

This argument remains the same if the predator exhibits type II behaviour, i.e. if a_{ij} is reduced when search time is lost due to prey handling[6][7]. We might represent such effects for example with the multispecies disc equation[8].

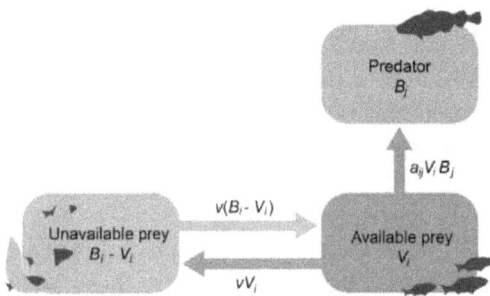

Figure 2. Simulation of flow between available (V_i) and unavailable ($B_i - V_i$) prey biomass in Ecosim. a_{ij} is the predator search rate for prey i, v is the exchange rate between the vulnerable and invulnerable state. Fast equilibrium between the two prey states implies $V_i = vB_i/ (2v + aB_j)$.[9]

Two specific models have been proposed for predicting changes in vulnerable prey densities V in foraging arenas[10]. The first or "continuous exchange" model[11] proposes that prey exchange between the vulnerable and invulnerable states at instantaneous rates v and v', so that V_i gains individuals at rate $v(B_i-V_i)$ and loses them at rates $v'V_i$ and aV_iB_j. This results in the rate equation

$$dV_i/dt = v(B_i - V_i) - v'\,V_i - a_{ij}V_i\,B_j \qquad (2)$$

If the vulnerability exchange and predation rates are high compared to overall rates of change of B_i and B_j, V_i is predicted to remain close to the moving equilibrium (with B_i and B_j) given by solving Eq. 2 with $dV_i/dt=0$:

$$V_i = vB_i/(v + v' + a_{ij}B_j) \qquad (3)$$

The second or "bout feeding" model proposes that prey (or predators) regularly (e.g., daily at dawn and dusk) enter the foraging arena for short temporal feeding bouts, depleting V_i exponentially during each bout such that the mean prey density seen by the predator during each bout of duration T is given by

$$V_i = vB_i(1 - e^{-aB_jT}/a_{ij}B_j) \qquad (4)$$

and initial vulnerable prey abundance vB_i is some fraction of the total prey population B_i. Note that both of these models imply two alternative ways to precisely define the phrase "limited food supply", found in ecological arguments[12], but generally lacks a formal definition. The supply of food may be defined as a temporal rate vN of food delivery to foraging arenas, or alternatively as the limited food density V that results from the balance of supply rate and removal processes.

An immediate and crucial prediction of models represented by Eq. 3 and Eq. 4 is that there can be strong negative effect of predator abundance B_j on vulnerable prey density V_i and feeding rate per predator Q/B_j, whether or not predators have any substantial impact on total prey abundance B_i[13]. Substituting Eq. 2 into Eq. 1 results in the "functional response" prediction

$$Q/B_j = a\,v\,B_i/(v + v' + aB_j) \qquad (5)$$

That is, the basic foraging arena models predict strong "ratio dependence" in the predation rate Q_j, with attendant consequences for predator-prey stability. Further, these models do not depend on specific assumptions about predator behaviour, such as interference or contest competition. Unlike models based on substituting B_j/B_i (prey per predator) ratios into functional response models they can be derived from fine-scale arguments about behaviour and spatial organization of interactions, and so are not subject to Abrams[14] very valid criticisms about the simplistic ratio formulations. Foraging arena models assert that competition between predators is intensified from the spatial restriction of interactions into arenas, however there is no one factor that restricts foraging activity, and restriction may arise result from factors such as prey and/or predator behaviours, or specific habitat requirements. Another basic prediction is that interaction rates Q_j can vary between "bottom-up" controlled and "top-down" controlled depending on v and a_{ij}. This is easiest to see with Eq. 2: If v is small and a_{ij} is large, Q_j approaches the "donor controlled" limiting rate vB_i as B_j increases; but as v increases, Q_j approaches the mass action rate $a_{ij}B_iB_j$.

The predictions from the foraging arena equations extend across a wide range of scales. Before describing these predictions in more detail, we find it important to demonstrate that the fundamental assumption of partitioning of prey into V_i and B_i-V_i components, with attendant exchange processes that can limit trophic interactions, is very widespread or potentially universal at least in aquatic ecosystems. Partitioning resulting from exchange processes implies a basic reversal of the idea that small proportions of prey may be in safe refuges so as to cause predation rates to have type III functional response form. Under the foraging arena assumption, it is far more common for the bulk of prey to be in refuges at any moment, particularly when exchange rates are low. Intense completion for resources within the foraging arena potentially results in increased foraging times by prey[15] as prey density increases, resulting in the type III form of the functional response due to changes in prey behaviour rather than predator behaviour.

Mechanisms that cause prey population partitioning and vulnerability exchange processes

It takes three to tango
If a small fish restricts its foraging to near hiding places, most become invulnerable at any moment to their predators.

A critical point about vulnerability exchange structures is that restriction in activity by any one species is likely to induce the exchange structure represented by Eq. 1 for at least two trophic linkages, between the species and its prey and between the species and its predator(s). Consider for example the simple food chain zooplankton $\rightarrow$ small fish $\rightarrow$ piscivore. If the small fish "chooses" to restrict its activities so as to forage only near hiding places, most of the small fish become invulnerable at any moment to piscivores. Likewise, then most of its zooplankton prey population becomes invulnerable to it at any moment. This "cascade" of foraging arena structures results in spatially limited interactions between predator and prey occurring on time scales of minutes/hours and at the spatial scale of meters (Figure 3), intensifying competition between predators when exchange processes limit the rate at which prey are replenished.

Flux prediction: Q=aVP

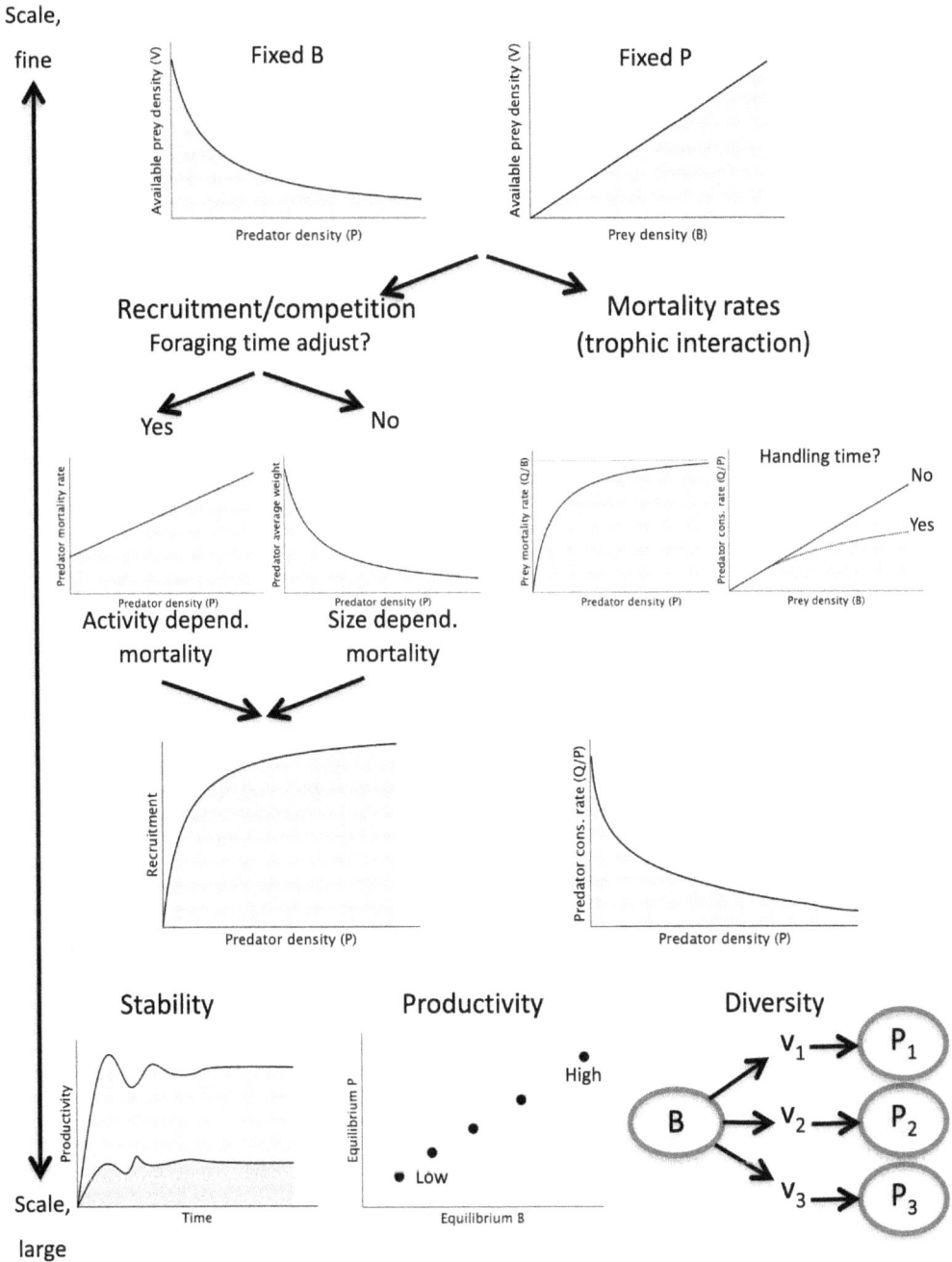

Figure 3. Foraging arena predictions across a range of space/time scales. The restriction of predator-prey interaction to "foraging arenas" results in a decreasing hyperbolic relationship between available prey density (V_i) and predator density (P_j) at fine space/time scales. Intra-specific competition within these arenas leads to the commonly observed Beverton-Holt recruitment relationship. For inter-specific

interactions, the exchange of prey into and out of these arenas limits predation mortality resulting in community stability. B_i is total prey biomass.

In the following section, we present a simple classification of behaviours that can lead to vulnerability exchange dynamics is presented. This classification is not complete or exhaustive, but it does cover a wide variety of trophic interactions in aquatic systems and demonstrates the broad applicability of foraging arena theory, (for relevant examples in the literature, see the source publication).

1. Arena structure caused by restricted spatial distribution of predators relative to prey

This category includes the original situation mentioned above, where the predator distribution covers only a small proportion of the area or volume occupied by prey organisms. But such restricted overlap can be caused by a variety of factors of which two appear to be particularly common. In all such cases, the vulnerability exchange rates v and v' are likely to have values determined mainly by physical transport (advection, diffusion) and random movement processes of the prey, and can be extremely low proportions of the overall prey population in physically large systems.

1.a. Restricted predator distribution in response to predation risk caused by its predators

The behaviour of post-larval juvenile fish is likely dominated by a need to reduce predation risk, and this is likely also the case for juveniles of mobile invertebrates. So far as we know from many examples, post-larvae move into highly restricted habitats (e.g., structure, schools) and spend relatively little time foraging. For most fish, increase in body size is associated with ontogenetic habitat shifts to use much larger foraging arenas and multiple habitat types.

Many mobile aquatic invertebrates exhibit strong vertical migration behaviours, apparently in response to predation risk but perhaps also as a way to manage metabolic costs or gain a horizontal transport advantage. Such behaviours result in temporally limited periods of overlap with prey, leading to diurnal foraging bouts and possibly localized prey depletion as represented by Eq. 4.

1.b. Restricted predator distribution caused by limited predator mobility or habitat requirements

Many "predators" have limited or no mobility, for example sessile invertebrates that filter-feed the water column above their resting site. Such restriction in vertical access to prey obviously creates a foraging arena exchange structure with algal and detritus "prey" distributed over the whole water column.

In some cases, apparently mobile predators still concentrate their activities in particular habitats even when not faced with obvious predation risk, perhaps as a way to manage energetic costs and/or places for ambushing prey. Many reef and demersal fish tend to hold and forage near bottom structure, even while taking mainly planktonic prey; one reason for this is that the ocean bottom acts as a trap to concentrate vertically migrating prey species. Such behaviour may be optimal under certain conditions and establishes an arena type structure.

2. Restricted prey distribution and/or activity

This category represents situations where predators may be widely distributed, but their prey show possibly severe restriction in spatial distribution and activity. Predators that exhibit type 1.a. behaviour above with respect to their prey are in turn expected to exhibit type 2 behaviour with respect to interactions with the species that prey on them.

2.a. Time allocation to safe/resting sites

This is the interesting case from an evolutionary perspective, where the same behaviours used to acquire resources (movement into foraging arenas to feed) cause some creatures to be the resources of other species (predation risk concentrated in the same arenas). Obviously such situations create trade-off relationships for which we can expect strong natural selection for optimized time allocation. It is difficult to generalize about the amount of time spent individuals under predation threat spend in refuge habitats. There is an indication that for juvenile fish, the optimum appears to typically be a small-time allocations to foraging particularly when foraging is restricted to crepuscular periods.

2.b. Vulnerability exchange associated with dispersal behaviours

The acquisition of resources is not the only behaviour that can expose organisms periodically to predation risk. Dispersal behaviours are also dangerous, and can occur for a wide variety of reasons (ontogenetic changes in habitat requirements or opportunities, response to locally high densities of competitors, movement to reproductive sites, etc.) Perhaps the most obvious example in aquatic systems is drift of benthic stream insects that spend most of their time in interstitial microhabitats where they are safe from most fish predation, but occasionally leave such sites to drift downstream. In this case, the V of Eq. 1 is literally the concentration of drifting (and emerging) insects, and the drift entry rate v can be limiting to potential abundance of stream predators like trout.

2.c. Vulnerability exchange caused by agonistic behaviours

Many aquatic organisms defend restricted resting or mating sites, and exhibit strong aggressive behaviours toward nearby conspecifics. In such cases, there can be strong density-dependent increase in agonistic activity with increasing density of conspecifics, leading to increased predation risk and density-dependent mortality at high densities.

2.d. High proportion of individual mass not accessible or digestible

Some predators take only parts of their prey without normally killing the prey. For example, browsing herbivores often select only particular plant parts that are physically accessible (not too high off the ground, not underground) or high in quality (seeds, leaves and active growth tips that are high in protein), leaving most of the prey growth/production system intact. In such cases, the v process represents prey body growth. Such dynamic structures are much more common in terrestrial than aquatic environments, but they do occur with grazers on macrophytes and macroalgae, and even with animal-animal interactions like fishes that nip at the siphons of buried molluscs or "graze" on parts of corals.

3. Spatial displacement of predators and prey by physical transport processes

It is common in aquatic ecosystems for production dynamics to be ordered in a physical flow pattern, where nutrient delivery at the head of the flow gives rise to primary production peaks downstream some distance (as primary producers are advected away from the nutrient source as they grow), and to secondary production peaks still further downstream as animals grow in response to the primary production as they are advected.

In such flow structures, smaller organisms may be able to partially control their downstream positions through counter-current movements (vertical migration, emergence and flying upstream). If these behaviours are not completely successful at bringing organisms to centers of prey abundance, such counter-current movements can result in organisms being concentrated in areas along the flow such that their food species appear to exhibit largely donor-controlled dynamics, i.e., to have concentration dynamics V with the same dominant terms (exchange in and out, predation loss) as in Eq. 1.

A similar concentration dynamic is observed when physical flow processes concentrate organism at frontal zones. These areas of higher food concentrations appear to be important foraging areas for higher trophic level organism such as sea birds or whales. In these structures, the concentration of production from a much wider areas establishes a foraging arena as organism exchange into frontal areas either through physical transport or directed movement.

Foraging arena predictions for a range of scales

A fundamental assumption of foraging arena theory is that predator-prey interactions occur at the scale of hours and meters through various behavioural and physical mechanisms potentially restricting prey exposure to predation and intensify competition between predators. This foraging arena formulation provides a structure that can be used to predicting observed states across a range of scale from the individual up to the ecosystem level.

At the scale of the individual, foraging arena theory can be invoked to explain the failure of fishes at least to consume nearly as much food as we would predict to be possible based on large-scale sampling of prey abundances. Back calculation of food intake rates from observed growth in the field, using laboratory-based bioenergetics models, indicates that fish typically feed at much lower rates than predicted from laboratory estimates of maximum ration. Fish biologists routinely encounter this phenomenon where a high proportion of the fish stomachs examined are empty. Foraging arena theory argues that the phenomenon is a symptom of evolutionary adaptation to predation risk, and involves two distinct and possibly interacting causes: spatial restriction in activity that leads to local prey depletion (low V) where foraging does take place, and/or temporal restriction in foraging activity also so as to reduce predation risk. Each of these causes can lead to empty stomachs or apparent reduced food intake. Suboptimal foraging has also been observed in the absence of predation though these observations have been for small individuals that may have restricted opportunity to select which areas to forage in. In addition individuals commonly stop or reduce feeding during spawning, brood rearing, and during migration, or may receive less than optimal ration due to territorial behaviours or dominance hierarchies.

The theory makes two broad predictions about what we should find when short-term (seasonal, annual) observations are collected across a range of predator densities. First, mean available food density per predator (V_i) should decrease in an inverse hyperbolic pattern as predator density B_j increases, with the first increments in predator abundance causing the greatest incremental decreases in available food density, whether or not there is any impact of B_j on the overall prey population B_i (Figure 3). This prediction is dependent on the exchange rates (v) and approaches a linear decrease in V_i with increasing B_j at higher exchange. Second, instantaneous prey mortality rate (Q_j/B_i) should increase in a hyperbolic pattern toward a maximum rate (v) as B_j increases, rather than simply being proportional to predator abundance B_j (Figure 3). When applied over longer time scales, this second prediction is the basic reason that predator-prey models based on foraging arena equations tend not to show cycles, even when handling time effects (reduction in predator search rate a with increasing B_i) are included in the predictions provided exchange rates (v) are low (right column of Figure 3).

On time scales of one to a few years, Walters and Korman[16] argue that the hyperbolic relationship between V_i and B_j, along with predator behaviour and predation risk, is likely to lead to the most commonly observed form of stock-recruitment relationship in fish populations, namely the flat-topped curve called the Beverton-Holt relationship (left column of Figure 3). Hundreds of empirical stock-recruitment relationships have been assembled for fish[17], and most of these show net recruitment to harvestable ages (typically 2-4 years) to be largely independent of parental spawning abundance or egg production. Such independence implies strong density-dependence in survival rates from egg to recruitment (else recruitment would on average be proportional to egg production, not independent of it). Beverton and Holt[18] showed that this pattern is expected if juveniles die off over time before recruitment according to a quadratic mortality model of the form $dB_j/dt = -(M0 + M1\ B_j)\ B_j$. Further, Walters and Korman[19] showed that exactly this linear relationship between instantaneous mortality rate $M0 + M1\ B_j$ is expected if (1) food density V available per B_j decreases as predicted by Eq. 2, juvenile fish adjust their daily foraging times so as to try and achieve a base growth rate needed to complete their ontogeny, and (3) mortality rate is proportional to time spent foraging.

Such predictions about individual and population scale patterns may help in interpreting some patterns in field data, but the really interesting predictions from foraging arena theory arise when models are developed for predicting impacts of changing trophic interactions in multispecies fisheries and whole aquatic food webs. Using the Ecopath mass-balance model to estimate initial abundances (B_i, B_j) and trophic flow rates (Q_j)

for a food web, changes in these abundances response to disturbances like fishing and changes in nutrient loading can be simulated over time.

It is easy to demonstrate that if we predict the changes in Q_j's using simple mass action rules ($Q_j = a_{ij}B_iB_j$, all species acting as though they were randomly mixed over the ecosystem), simulated competition and predation effects quickly result in substantial loss in food web structure. Such model pathologies only become worse when we include more realistic, type II functional response effects representing limitation on predator feeding rates due to handling times and adjustments in foraging times to achieve target food consumption rates; the typical result is to predict at least some predator-prey oscillations, along with "paradox of enrichment" effects (increasing dynamic instability as simulated primary productivity is increased).

When food web models like EwE-Ecosim are used to predict effects of dynamic changes in predator-prey interaction rates Q_j using the foraging arena vulnerability exchange equations (Eq. 1 to Eq. 5), there is a dramatic reversal of the difficulties encountered with models based on simple mass action interaction rates. Models with low vulnerability exchange rates (v's) routinely make four key predictions that are difficult to obtain with simplified mass action models:

1. Predator-prey cycles should be rare in aquatic ecosystems, and no paradox of enrichment (instability at high productivity) should occur along spatial or temporal gradients in primary productivity.

2. Trophic cascades[20] should be common at least in simpler aquatic ecosystems

3. The Gauss "competitive exclusion principle"[21](Hardin, 1960) should fail.

4. In harvested systems, surplus production of predators should be created by immediate compensatory responses to increased per-capita food density (availability) in foraging arenas.

Assessment of vulnerability exchange rates for ecosystem management models

There is a clear need for quantitative models to evaluate the various trade-offs involved in aquatic ecosystem management, so as to provide advice that can at least rank the relative impact of management options and to expose critical uncertainties that may trigger precautionary or experimental management policies. We doubt that any natural historian who has looked closely at spatial and temporal organization of aquatic trophic interactions would doubt the need to represent such interactions as being restricted to at least some degree to what we have called foraging arenas, whether or not such arenas can be precisely defined and measured under field conditions. It is likely that interactions between predators and prey are occurring at the scale of hours and meters. But in practice, there is a huge gulf between knowing that interaction rates are likely to be restricted to some degree by vulnerability exchange rates (v), versus being able to quantify such rates well enough to say whether they are low enough to require abandonment of simpler mass-action predictions of interaction rates, and to make useful predictions about compensatory responses (surplus production) to various disturbance regimes.

A variety of approaches have been tried for estimating vulnerability exchange rates from field data. None of these has been fully satisfactory, at least partly because arena structures in the field are spatially and temporally complex; indeed, one reason to call the foraging arena arguments a "theory" is that arena structures are "theoretical entities" that are practically difficult or impossible to directly observe[22].

Three main methods have been used to provide estimates of apparent v's using field data, and a fourth is under development.

1. Direct assessment of exchange rates for spatially simple arena structures

2. Empirical relationships between prey mortality rates and predator abundances

3. Fitting ecosystem models to time series data

4. Using complex individual-based spatial models

See the source[23] for details about these.

The main modeling argument for assuming mass action in predictions of predator-prey and food web interaction effects has never been that such a simplistic assumption is warranted based on field data; rather, the use of such models can be justified mainly because of analytical and computational tractability, i.e., the notion that robust and general predictions cannot be easily derived for more realistic models. The models of foraging arena theory, and associated software like EwE-Ecosim for examining dynamic scenarios, largely eliminate such excuses. We assert that the issue now for ecosystem modeling is not whether to bother including vulnerability exchange effects in trophic interaction predictions (it is plainly unwise to ignore them), but rather how to estimate exchange rates and their impacts.

Attribution

The chapter is adapted from Ahrens, R.N.M., Walters, C.J. and Christensen, V. (2012), Foraging arena theory. Fish and Fisheries, 13: 41-59. https://doi.org/10.1111/j.1467-2979.2011.00432.x with permission from John Wiley and Sons, license numbers 5676200521292 and 5676200679690. Please cite the original source instead of this chapter.

Media Attributions

- Figure 1 from Ahrens et al. 2012
- From Bentley et al. 2024 Figure 1
- Figure 2 from Ahrens et al. 2012

Notes

1. Walters, C.J., Juanes, F. 1993. Recruitment limitation as a consequence of natural-selection for use of restricted feeding habitats and predation risk-taking by juvenile fishes. Canadian Journal of Fisheries and Aquatic Sciences 50, 2058-2070. https://doi.org/10.1139/f93-22

2. Walters, C., Christensen, V., Pauly, D. 1997. Structuring dynamic models of exploited ecosystems from trophic mass-balance assessments. Reviews in Fish Biology and Fisheries 7, 139-172. https://doi.org/10.1023/A:1018479526149

3. Walters, C., Korman, J. 1999. Linking recruitment to trophic factors: revisiting the Beverton-Holt recruitment model from a life history and multispecies perspective. Reviews in Fish Biology and Fisheries 9, 187-202. https://doi.org/10.1023/A:1008991021305.

4. Review in Walters, C.J., Martell, S.J. 2004. Fisheries ecology and management, Vol., Princeton University Press, Princeton, New Jersey

5. Ahrens, R.N.M., Walters, C.J. and Christensen, V. (2012), Foraging arena theory. Fish and Fisheries, 13: 41-59. https://doi.org/10.1111/j.1467-2979.2011.00432.x

6. Holling, C.S. (1959a) The components of predation as revealed by a study of small-mammal predation of the European pine sawfly. The Canadian Entomologist 91, 293-320. https://doi.org/10.4039/Ent91293-5

7. Holling, C.S. (1959b) Some characteristics of simple types of predation and parasitism. The Canadian Entomologist 91, 385–398. https://doi.org/10.4039/Ent91385-7

8. May, R.M. (1973) Stability and complexity in model ecosystems, Monographs in Population Biology, Vol. 6, Princeton University Press, Princeton, New Jersey.

9. Based on Walters, C., V. Christensen and D. Pauly. 1997. *Op. cit.*

10. Walters, C., Christensen, V. (2007) Adding realism to foraging arena predictions of trophic flow rates in Ecosim ecosystem models: Shared foraging arenas and bout feeding. Ecological Modelling 209, 342-350. https://doi.org/10.1016/j.ecolmodel.2007.06.025

11. Walters et al., 1997. *op. cit.*

12. e.g., Abrams, P.A., Ginzburg, L.R. (2000) The nature of predation: prey dependent, ratio dependent or neither? Trends in Ecology & Evolution 15, 337-341. https://doi.org/10.1016/S0169-5347(00)01908-X

13. as suggested in Abrams and Ginzburg, 2000, *op. cit.*

14. Abrams, P.A. (1994) The fallacies of ratio-dependent predation. Ecology 75, 1842-1850. https://doi.org/10.2307/1939644

15. see Walters and Juanes, 1993. *op. cit.*

16. Walters and Korman 1999. *op. cit.*

17. RAM Legacy Stock Assessment Database, records available at https://zenodo.org/records/7814638

18. Beverton, R.J.H., Holt, S.J. (1957) On the dynamics of exploited fish populations. U.K. Ministry of Agriculture, Fisheries and Food, Fisheries Investigations Series 2 19, 533. https://link.springer.com/book/10.1007/978-94-011-2106-4

19. Walters and Korman 1999. *op. cit.*

20. Carpenter, S.R. and Kitchell, J.F. (1993) The Trophic Cascade in Lakes. Cambridge University Press, Cambridge. http://dx.doi.org/10.1017/CBO9780511525513

21. Hardin, G.J. (1960). The competitive exclusion principle. Science, 131 3409, 1292-7. https://www.science.org/doi/10.1126/science.131.3409.1292

22. Maxwell, G. (1962) The ontological status of theoretical entitites. In: Minnesota Studies in the Philosophy of Science, vol. III: Scientific Explanation, Space, and Time. (Eds. H. Feigl, G. Maxwell), University of Minnesota Press, Minneapolis, MN, pp. 3-27. https://conservancy.umn.edu/bitstream/handle/11299/184634/3-01_Maxwell.pdf?sequence=1

23. Ahrens et al. 2012. *op. cit.*

15.

A primer on dynamic modelling

Dynamic modelling is about prediction over time, for which there are two approaches,

- Trend projections from past observations
- Prediction based on "rules for change" (how variables interact)

The dynamic modelling in EwE is based rules for change of two types,

- Tautologies that organize components of change, e.g., *Next population = population new + births – deaths*
- Functional relationships that propose testable hypotheses about how the components vary, e.g., *Births = constant x population size.*

The rules for change include two types of parameters,

- "State variables" that change over time, e.g, N_t the number in a population at time t.
- "Parameters" that are assumed constant over time, e.g., a birth rate.

There are two basic ways to represent rules for change,

- Difference equations, e.g.,

$$N_{t+1} = N_t + bN_t - mN_t \tag{1}$$

  where b = birth rate and m = death rate
- Differential equations, e.g.,

$$dN/dt = bN_t - mN_t \tag{2}$$

For complex continuous time models, such differential equations can be solved by simple numerical stepping approximations. For small time steps, we may use the one-step "Euler" method,

$$N_{t+dt} = N_t + X \cdot dt \tag{3}$$

where $X = dN/dt$ at time t. The Euler method thus predicts change using only the rate at time t. It can give very poor results for rate equations where there are strong positive feedbacks, i.e. when X can increase rapidly as N increases.

For implementing dynamic model calculations in spreadsheets, it is better though to use the more precise, two-step "Adams-Bashforth" method,

$$N_{t+dt} = N_t + [3X - Y]/2 \cdot dt \qquad (4)$$

where X is as in Eq. 3 and $Y = dN/dt$ for the previous time step, time t-dt. The Adams-Bashforth is used for the tutorial on predator-prey models (included in the web and pdf-versions of this book).

For numerical integration in computer languages like R and VB.Net, it is better though to use 4th order Runge-Kutta integration, especially if the model includes state variables that change at very different rates (both "fast" and "slow" dynamic change). The dynamic rate equations in the EwE software are solved using this more accurate integration method, which approximates the curve between two points by at 4th degree polynomial.

16.

An introduction to Ecosim

Ecosim provides a dynamic simulation capability at the ecosystem level, with key initial parameters inherited from the base Ecopath model.

The key computational aspects are in summary form,

- Use of mass-balance results (from Ecopath) for parameter estimation;

- Variable speed splitting enables efficient modelling of the dynamics of both "fast" (e.g., phytoplankton) and "slow" groups (e.g., whales);

- Effects of micro-scale behaviours on macro-scale rates: top-down vs. bottom-up control incorporated explicitly.

- Includes biomass and size structure dynamics for key ecosystem groups, using a mix of differential and difference equations. As part of this EwE incorporates:

- Multi-stanza life stage structure by monthly cohorts, density- and risk-dependent growth – described in the Age-structured dynamics chapter;

- Stock-recruitment relationship as "emergent" property of competition/predation interactions of juveniles.

Ecosim uses a system of differential equations that expresses biomass flux rates among pools as a function of time varying biomass and harvest rates, (for equations see Walters et al., 1997[1]; Walters et al., 2000[2]; Christensen and Walters, 2004[3]). Predator prey interactions are moderated by prey behaviour to limit exposure to predation, such that biomass flux patterns can show either bottom-up or top down (trophic cascade) control. By doing repeated simulations, Ecosim allows for the fitting of predicted biomasses to time series data.

The simplest, default version of Ecosim represents biomass dynamics using a series of coupled differential equations. The equations are of the basic form:

$$\frac{dB_i}{dt} = g_i \sum_{i=1}^{n} Q_{ij} - \sum_{j=1}^{n} Q_{ji} + I_i - (F_i + e_i + M0_i)B_i \tag{1}$$

where dB_i/dt represents the growth rate during the time interval dt of group i in terms of its biomass, B_i, g_i is the net growth efficiency (production/consumption ratio), $M0_i$ the non-predation ("other") natural mortality rate, F_i is fishing mortality rate, e_i is emigration rate, I_i is immigration rate, (and $e_i \cdot B_i - I_i$ is the net migration rate). The two summations estimate consumption rates, the first expressing the total consumption by group i, and the second the predation by all predators j on the prey group.

Ecopath is used to provide the initial (t=0) biomasses, and some of the rate parameters (like MO). Ecosim parameters for the flow or consumption rates Q_{ij} are set partly from Ecopath base estimates of those flows, with addition information needed to represent how the flow rates vary with biomasses and other circumstances.

The consumption rates, Q_{ji} and Q_{ij}, represent consumption by group j on i and by i on j, respectively, and are calculated based on foraging arena theory, where B_i's are divided into vulnerable and invulnerable components[4], and it is the transfer rate (v_{ij}) between these two components that determines if control is top-down (i.e., Lotka-Volterra), bottom-up (i.e., donor-driven), or of an intermediate type. See the vulnerability multiplier chapter.

The set of differential equations is solved in Ecosim using a 4th order Runge-Kutta routine (see the A primer on dynamic modelling chapter). For groups like phytoplankton and small zooplankton that turn over (have P/B) greater than 10 and are likely to exhibit boom-bust dynamics on time scales shorter than one month, the numerical integration prediction is replaced with a prediction based on the equilibrium of the Ecosim rate equation of the likely average over the month.

Attribution This chapter is in part adapted from the unpublished EwE User Guide: Christensen V, C Walters, D Pauly, R Forrest. Ecopath with Ecosim. User Guide. November 2008.

Notes

1. Walters, C., V. Christensen and D. Pauly. 1997. Structuring dynamic models of exploited ecosystems from trophic mass-balance assessments. Reviews in Fish Biology and Fisheries 7:139-172.

2. Walters, C.J., J.F. Kitchell, V. Christensen and D. Pauly. 2000. Representing density dependent consequences of life history strategies in aquatic ecosystems: Ecosim II. Ecosystems 3: 70-83.

3. Christensen, V. and C. J. Walters. 2004. Ecopath with Ecosim: methods, capabilities and limitations. Ecol. Model. 172:109-139

4. Figure 1 in Walters, C., V. Christensen and D. Pauly. 1997. Structuring dynamic models of exploited ecosystems from trophic mass-balance assessments. Reviews in Fish Biology and Fisheries 7:139-172. https://doi.org/10.1023/A:1018479526149

17.

Predicting consumption

This is what Ecosim (and all other dynamic ecosystem models) really is about. If the number of consumers change over time, how much do they eat? How does consumption change with population density?

All ecosystem models predict consumption (Q_{ij}) changes based on a variant of Lotka-Volterra dynamics (see chapter), including Ecosim which uses simple Lotka-Volterra or "mass action" assumptions for prediction of consumption rates. But importantly, the assumption is modified to consider "foraging arena" properties so that the flow rates depend on abundance of vulnerable prey rather than total prey abundance. In the foraging arena model structure, prey can be in states that are or are not vulnerable to predation, for instance by hiding, (e.g., in crevices in reefs, inside a school, where predators don't go) when not feeding, and only being subject to predation when having left their shelter to feed. (see chapter).

In the original Ecosim formulations[1] [2]) the foraging arena consumption rate for a given predator i feeding on a prey j was predicted as,

$$Q_{ij} = \frac{a_{ij}\, v_{ij}\, B_i\, B_j}{2v_{ij} + a_{ij}\, B_j} \qquad (1)$$

where, a_{ij} is the effective search rate for predator j feeding on a prey i, v_{ij} base vulnerability expressing the rate with which prey move between being vulnerable and not vulnerable, B_i prey biomass, and B_j predator abundance (for multi-stanza groups, B_j in this calculation is replaced by an estimate of the area swept by organisms of varying sizes, summed over ages within each stanza).

For discussion about the relationship between top-down vs bottom-up and carrying capacity, see the Density dependence chapter.

The model as implemented implies that "top-down vs. bottom-up" control is in fact a continuum, where low v's implies bottom-up and high v's top-down control.

Experience with Ecosim has led to a more elaborate expression to describe how consumption may vary with a variety of factors:

$$Q_{ij} = \frac{v_{ij}\, a_{ij}\, B_i\, B_j\, T_i\, T_j\, S_{ij} M_{ij}/D_j}{v_{ij} + v_{ij}\, T_i\, M_{ij} + a_{ij}\, M_{ij}\, B_j\, S_{ij}\, T_j/D_j/A} \cdot f(Env_t) \tag{2}$$

where, T_i represents prey relative feeding time, T_j predator relative feeding time, S_{ij} user-defined seasonal or long term forcing effects, M_{ij} mediation forcing effects, A is foraging arena size, $f(Env_t)$ is an environmental response function that impacting the size of the foraging arena to account for external drivers, which may change over time[3], and D_j represents effects of handling time as a limit to consumption rate ($1/D_j$ is proportion of time spent feeding):

$$D_j = 1 + h_j \sum_k a_{kj} V_k T_k M_{kj} \tag{3}$$

where h_j is the predator handling time and V_k is the vulnerable density of prey type k to predator j (V_k is estimated numerically in the Ecosim code).

The food consumption prediction relationship in the second equation above contains two parameters that directly influence the time spent feeding and the predation risk that feeding may entail: a_{ij} and v'_{ij}. To model possible linked changes in these parameters with changes in food availability over time (t) as measured by per biomass food intake rate $c_{i,t} = Q_{i,t} / B_{i,t}$ we need to specify how changes in $c_{i,t}$ will influence at least relative time spent foraging.

> For multi-stanza groups, B_i is replaced by a sum over ages of numbers at age times body weights to the ⅔ power.

Denoting the relative time spent foraging as $T_{i,t}$, measured such that the rate of effective search during any model time step t can be predicted as $a_{ji,t} = T_{i,t}\, a_{ij}$ for each prey type i that j eats, we may (optionally) assume that time spent vulnerable to predation, as measured by v'_{ij} for all predators j on i, is inversely related to $T_{i,t}$, i.e., $v'_{ij,t} = v'_{ij} / T_{i,t}$. An alternative structure that gives similar results is to leave the a_{ij} constant, while varying the v_{ij} by setting $v_{ij,t} = T_{j,t} \cdot v_{ij}$ in the numerator of Eq. 2 and $v_{ij,t} = T_{i,t} \cdot v_{ij}$ in the denominator.

For convenience in estimating the a_{ij} and v'_{ij} parameters, we scale $T_{i,t}$ so that $T_{i,0} = 1$, and $v'_{ij} = v_{ij}$. Using these scaling conventions, the key issue then becomes how to functionally relate $T_{i,t}$ to food intake rate $c_{i,t}$ so as to represent the hypothesis that animals with lots of food available will simply spend less time foraging, rather than increase food intake rates.

In Ecosim, a simple functional form for $T_{i,t}$ is implemented that will result in near constant feeding rates, but changing time at risk to predation, in situations where rate of effective search a_{ij} is the main factor limiting food consumption rather than prey behaviour as measured by v_{ji}. This is implemented in form of the relationship,

$$T_{i,t} = T_{i,t-1}(1 - \alpha + \alpha \frac{Q_{opt,i,t}}{Q_{i,t-1}}) \tag{5}$$

where, a is a feeding time adjustment rate [0, 1]and Q_{opt} is the Ecopath base consumption rate per biomass (QB) for group j . This calculation is subject to a user-defined maximum relative foraging time for each predator, and the result of that upper limit is for the predator functional response to be approximately the Holling Type I (rectilinear, see Holling functional response chapter) form with steepness proportional to the maximum relative foraging time.

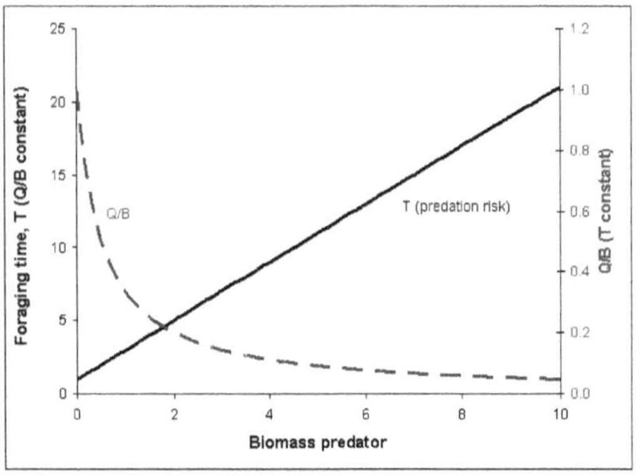

The relationship between foraging time, consumption and predator biomass (when adjustment rate a is assumed nonzero) is illustrated in Figure 1.

Figure 1. Relationship between relative foraging time (T), Q/B and predator biomass. If Q/B is held constant the foraging time (and hence predation risk) is a linear function of the predator biomass (solid line). If T is held constant the Q/B will decrease asymptotically with predator biomass (stippled line). The predation risk is assumed proportional to the relative foraging time.

Attribution: This chapter is in part adapted from the unpublished EwE User Guide: Christensen V, C Walters, D Pauly, R Forrest. Ecopath with Ecosim. User Guide. November 2008.

Media Attributions

- Figure 1 from the 2008 EwE User Guide

Notes

1. Walters, C., V. Christensen and D. Pauly. 1997. Structuring dynamic models of exploited ecosystems from trophic mass-balance assessments. Reviews in Fish Biology and Fisheries 7:139-172.

2. Walters, C.J., J.F. Kitchell, V. Christensen and D. Pauly. 2000. Representing density dependent consequences of life history strategies in aquatic ecosystems: Ecosim II. Ecosystems 3: 70-83.

3. Christensen, V, M Coll, J Steenbeek, J Buszowski, D Chagaris, and CJ Walters. 2014. Representing variable habitat quality in a spatial food web model. Ecosystems 17(8): 1397-1412. http://www.jstor.org/stable/43678116

18.

Age-structured dynamics

The default approach in EwE is to model functional groups as biomass pools that in Ecosim have very simple dynamics. For such, Ecosim – as described in previous chapters – solves a set of differential equations for biomass rate of change of the form,

$$dB_i/dt = eQ_i(t) - Z_i(t)\, B_i \tag{1}$$

This simple representation does not allow for modelling species with complex trophic ontogeny or size-age dependent fishery impacts. Such groups can, however, be designated age-structured life-history stanzas within single-species populations. In such cases, the Ecosim differential equation solution for biomass change (Eq. 1) is replaced by a monthly-difference equation system, with full monthly age-structured accounting for population age and size structure.

With this approach, Ecosim can be used to simulate monthly changes in numbers and relative body weights of monthly age cohorts of species with complex trophic and fisheries impact ontogeny. For this, the start is to split any species into an arbitrary number of age (in months) "stanzas" as described earlier in the multi-stanza life history chapter. For such groups, prey preferences and vulnerability to predators (and fisheries) is then treated as constant over the months of age included within each stanza. Stanza age breaks can represent both ontogenetic shifts in habitat and diet and changes in vulnerability to bycatch and retention fisheries.

Ecosim differential equation representation for biomass change is replaced by a monthly-difference equation system, with full age-structured accounting for population age and size structure at monthly age increments. The basic accounting relationships are

$$N_{a+1,t+1} = N_{a,t}\, \exp(-Z_{s,t}/12) \tag{4}$$

$$W_{a+1,t+1} = \alpha_a q_{a,t} + \rho W_{a,t} \tag{5}$$

$$B_{s,t} = \sum_{a=a1(s)}^{a2(s)} N_{a,t} W_{a,t} \tag{6}$$

Where, $N_{a,t}$ is the number of age a (in months) animals in calendar month t, $W_{a,t}$ is the mean body weight of age a animals in month t, and $B_{s,t}$ is the biomass of stanza s, defined as the mass (numbers × weight) of animals aged $a1(s)$ through $a2(s)$ months. $Z_{s,t}$ is the total mortality rate of stanza s animals, defined the same way on the basis of fishing and consumption as for other model biomass groups i as $Z_{s,t} = M_{os} + \Sigma_f F_{sf} + \Sigma_j Q_{sj}/B_s$. All animals in stanza s are treated as having the same predation risk and vulnerability to fishing. The aggregated bioenergetics parameters a_a and r are calculated to make body growth follow a von Bertalanffy growth curve (with length-weight power 3.0) with user-defined metabolic parameter K. Exact von Bertalanffy growth occurs when predicted per-capita food intake $q_{a,t}$ is equal to a base food intake rate that is calculated from the consumption per biomass parameter (Q_s/B_s) provided by the user for each stanza. The metabolic parameter r, which equals $\exp(-3K/12)$, is based on the assumption that metabolism is proportional to body weight[1]. Actual or realized food intake $q_{s,t}$ at each time step is calculated from the total predicted food-intake rate for the stanza $(Q_{s,t})$ as $q_{s,t} = Q_{s,t} w_{a,t}^{2/3}/P_{s,t}$, where $P_{s,t}$ is the relative total area searched for food by stanza s animals and is computed as $P_{s,t} = \Sigma_a N_{a,t} w_{a,t}^{2/3}$. For foraging-arena food-intake and predation-rate calculations involving stanzas, $P_{s,t}$ is used instead of B_s as the predictor of total area or volume searched for food per unit time. The assumption that area searched and food intake vary as the ⅔ power of weight (i.e., as the square of body length) is a basic assumption that also underlies the derivation of the von Bertalanffy growth function.

For notational simplicity, Eqs. 4–6 above are presented without a species index. Typical Ecosim models developed to date have included multistanza accounting for 2–10 species, each divided into 2–5 stanzas that capture basic ontogenetic changes in diet, predation risk, and vulnerability to fishing. The first age for stanza 1 is always set to $a1(1) = 0$ (hatching), and $a2(1)$ is often set to 3–6 months of age to represent the larval and early juvenile periods separately. Then $a2(2)$ is often set at 12–24 months (to represent older juveniles), and additional stanza breaks are set at key ages like maturity and first vulnerability to fishing.

Initial numbers entering the first stanza for multistanza species s each month are assumed to be proportional to total egg production, and egg production is assumed to be proportional to body weight minus a weight at maturity $W_{s,mat}$. That is,

$$N_{1,t} = k_s \sum_a N_{a,t}[W_{a,t} - W_{s,mat}] \tag{7}$$

The effective fecundity parameter, k_s is calculated from initial numbers, $N_{1,0}$, and these initial numbers are calculated in turn from Ecopath input values of biomass for one "leading" stanza for each species s, along with initial survivorships to age calculated from initial Ecopath input values of $Z_{a,0}$. For these calculations, relative body weights, $W_{a,0}$, are set initially to the von Bertalanffy prediction, and weight is assumed to vary as the cube of length, as $W_{a,0} = (1 - e^{-Ksa})^3$.

Egg production is allowed to vary seasonally or over long-term through an input forcing function. If an egg production curve is defined, the egg production term is multiplied according to the forcing function. Note that this age-0 recruitment formulation for newly entering animals proportional to egg production does not explicitly account for density dependence in early mortality rates (i.e., an explicit stock-recruitment function is not used). Density-dependent effects occur through (1) impacts of animal density on food consumption, growth, and fecundity (a time-lagged effect that can result in violent population cycles) and, more importantly and commonly, (2) density dependence in $Z_{s,t}$ caused by foraging-time adjustments in the Ecosim foraging-arena model for $Q_{s,t}$. Foraging-time adjustments typically result in emergent stock-recruitment relationships of Beverton-Holt form.[2][3]

The Ecosim multistanza model has been fitted to many time series of population abundances that were reconstructed from single-species age-structure data by methods like VPA[4] and stock-reduction analysis[5]. Species fitted range from tunas to groupers to small pelagics like menhaden. For large, relatively long-lived species (piscivores, benthivores), behavior of the multistanza population model is typically indistinguishable from those of other age-structured models commonly used for stock assessment. For small-bodied species subject to high and temporally varying predation-mortality rates (e.g., small tunas, herrings, menhaden), Ecosim can sometimes capture effects such as relative stability of Z as F increases (decreases in M with increasing F) that are typically missed by singlespecies models that assume stable natural mortality rate M[6].

On entry to Ecosim from Ecopath, the stanza age-size distribution information (l_a, W_a) is passed along and is used to initialize a fully size-age structured simulation for the multistanza populations. That is, for each monthly time step in Ecosim, numbers at monthly ages $N_{a,t}$ and body weights $W_{a,t}$ are updated for ages up to the 90% maximum body weight age (older, slow growing animals are accounted in an "accumulator" age group). The body growth $W_{a,t}$ calculations (Eq 5) are parameterized so as to follow von Bertalanffy growth curves[7], with growth rates dependent on body size and (size- and time-varying) food consumption rates.

Because a biomass-age pattern (and food consumption–age pattern proportional to $w^{2/3}$) like that in Figure 1 must be satisfied once stanza-specific base Z's have been specified for every stanza, initial biomass B_{i0} and food consumption per biomass, Q_{i0}/B_{i0}, can only be entered for one stanza (biomass pool i) for each multi-stanza species s. Then B and Q/B are calculated for the other stanzas from the relative (per recruit) biomass and food consumption rates summed over ages in those stanzas.

At this point, the Ecosim age structured dynamics behave pretty much the same as standard age-structured models for single species assessments, but with the important exception that such models typically include an explicit stock-recruitment relationship. Instead of using such a relationship, Ecosim generates an "emergent" compensatory relationship through assumptions related to the foraging arena equations; see the following Recruitment and compensation chapter for details.

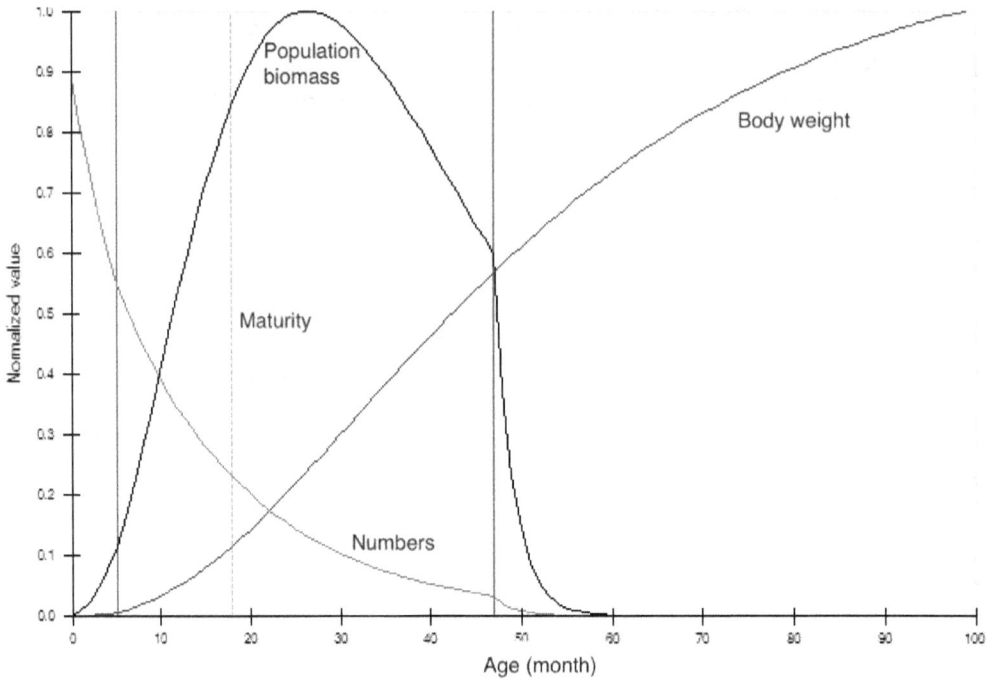

Figure 1. Typical changes in individual biomass and population numbers and biomass with age for a multi-stanza population in Ecosim. In this example, von Bertalanffy $K = 0.4$ year^{-1}, and the three stanzas marked by the outer vertical lines have the following total mortality rates, Z: 1.2 year^{-1} for age 0–5 months, 0.8 year^{-1} for age 6–47 months, and 6.0 year^{-1} for ages 48+ months (moribund salmon). The dotted vertical line indicates age-at-maturity, which can be overruled by setting a spawning/not-spawning variable for each stanza.

Attribution The chapter is in part adapted from Walters et al. 2008[8] and 2010[9] both in the *Bulletin of Marine Science,* which permits authors to use figures, tables, and brief excerpts in scientific and educational works provided that the source is acknowledged and the use is non-commercial.

Notes

1. Essington, T. E., J. F. Kitchell, and C. J. Walters. 2001. The von Bertalanffy growth function, bioenergetics, and the consumption rates of fish. Can. J. Fish. Aquat. Sci. 28: 2129–2138. https://doi.org/10.1139/f01-151

2. Walters C and J. Korman. 1999. Revisiting the Beverton-Holt recruitment model from a life history and multispecies perspective. Rev. Fish Biol. Fish. 9: 187–202. https://doi.org/10.1023/A:1008991021305

3. Walters C and S. J. D. Martell. 2004. Fisheries Ecology and Management. Princeton Univ. Press, Princeton. 399 p.

4. Sparre, P. 1991. An introduction to multispecies virtual analysis. ICES Mar. Sci. Symp. 193: 12–21.

5. Walters CJ, Martell SJD, and Korman J. 2006. A stochastic approach to stock reduction analysis. Can. J. Fish. Aquat. Sci. 63: 212–223. https://doi.org/10.1139/f05-213

6. see, e.g., Walters, C, Martell, SJD, Christensen, V, and Mahmoudi, B. 2008. An Ecosim model for exploring ecosystem management options for the Gulf of Mexico: implications of including multistanza life history models for policy predictions. Bull. Mar. Sci. 83(1): 251-271

7. Von Bertalanffy, *op. cit.*

8. Walters, C, Martell, SJD, Christensen, V, and Mahmoudi, B. 2008. An Ecosim model for exploring ecosystem management options for the Gulf of Mexico: implications of including multistanza life history models for policy predictions. Bull. Mar. Sci. 83(1): 251-271

9. Walters, C., Christensen V, Walters W, Rose K. 2010. Representation of multi-stanza life histories in Ecospace models for spatial organization of ecosystem trophic interaction patterns. Bull. Mar. Sci. 86(2):439-459

19.

Recruitment and compensation

Compensatory mechanisms

Sustaining fisheries yield when fishing reduces stock size depends on the existence of compensatory improvements in per capita recruitment, growth, and/or natural mortality rates. Ecosim allows users to represent a variety of specific hypotheses about compensatory mechanisms. Broadly, these mechanisms fall in two categories:

- direct – changes caused over short time scales (order one year) by changes in behaviour of organisms, whether or not there is an ecosystem-scale change due to fishing; and

- indirect – changes over longer time scales due to ecosystem-scale responses such as increased prey densities and/or reduced predator densities. Usually we find the direct effects to be most important in explaining historical response data. In the next three sections we describe how to generate alternative models or hypotheses about direct compensatory responses; these hypotheses fall in three obvious categories: recruitment, growth and natural mortality.

Using Ecosim to study compensation in recruitment relationships

The multi-stanza representation of juvenile and adult biomasses was originally included in Ecosim to allow representation of trophic ontogeny (big differences in diet between juveniles and adults). To implement this representation, we found that it was necessary to include population numbers and age structure, at least for juveniles, so as to prevent "impossible" dynamics such as elimination of juvenile biomass by competition/predation or fishing without attendant impact on adult abundance (graduation from juvenile to adult pools cannot be well represented just as a biomass "flow").

When we elected to include age-structured dynamics, we in effect created a requirement to think carefully about the dynamics of compensatory processes that have traditionally been studied in terms of the "stock-recruitment" concept and relationships. To credibly describe the dynamics of multi-stanza populations, Ecosim parameters for multi-stanza juvenile stages usually need to be set so as to produce an 'emergent" stock- recruitment relationship that is at least qualitatively similar to the many, many relationships for which we now have empirical data (see the RAM Legacy Stock Assessment Database). In most cases, these relationships are "flat" over a wide range of spawning stock size, implying there must generally be strong compensatory increase in juvenile survival rate as spawning stock declines (otherwise less eggs would mean less recruits on average, no matter how variable the survival rate might be).

When creating multi-stanza dynamics, be careful in setting model parameters that define/create compensatory effects. This begins with the Ecopath input parameters; in order for the juvenile dynamics to display compensatory mortality changes, at least two conditions are needed or helpful:

- the juvenile group(s) must have relatively high *P/B*, i.e. high total mortality rate (see Multi-stanza life history chapter);

- the juvenile group(s) must have either relatively high *EE* (so that most mortality is accounted for as predation effects within the model) or else the user must specify a high (near 1.0) value in the *Ecosim > Input > Group info* form entry for the juvenile group's *Proportion of other mortality sensitive to changes in feeding time* column.

Compensatory effects can be increased (the recruitment relationship is flat over a wider range of adult stock sizes, with a steeper slope of recruitment curve near the origin) by:

- Limiting the availability of prey to juveniles (forcing juveniles to use small foraging arenas for feeding) by setting all elements of the *Ecosim vulnerability multiplier* form column for the juveniles to a low value (1.1-2.0); or

- Setting a higher value for the juvenile group's *Feeding time adjustment rate* parameter (*Ecosim > Input > Group info* form), which causes the effective time exposed to predation while feeding to drop directly with decreasing juvenile abundance (i.e., simulates the possibility that when juveniles are less abundant, remaining ones may be able to forage "safely" only in refuge sites without exposing themselves to predation risk). This option should preferably be used if you are fairly sure from field natural history observation that the juveniles do in fact restrict their distribution to safe habitats when at very low abundance.

It is especially important to test alternative values for the vulnerability of prey to juveniles. If the vulnerability multiplier is too high, the Ecosim emergent stock-recruitment relationship is likely to look almost like a straight line out of the origin, i.e. without compensatory effect. If the vulnerability multiplier is too low, the relationship may develop a "spurious" dome-shape.

In Ecosim multi-stanza groups, the group that is displayed on the *Ecosim > Output > Stock recruitment* plot (*S/R*) is always the oldest stanza. The stock-recruitment relationship between this stage and each of the younger stages separately is calculated and displayed on the *S/R* form. This may cause issues when the oldest group is a "senescent" group as is often done for modelling Pacific salmon, (which die after spawning).

Compensatory growth

Compensatory growth rate responses are modelled by setting the *feeding time adjustment rate (Ecosim > Input > Group info form)* to zero, so that simulated Q/B is allowed to vary with the group biomass (non-zero feeding time adjustment results in simulated organisms trying to maintain Ecopath base Q/B by varying relative feeding time). Net production is assumed proportional (growth efficiency) to Q/B, whether or not this production is due to recruitment or growth. The Q/B increase with decreasing pool biomass is increased by decreasing vulnerability of prey to the pool (*Ecosim > Input > Vulnerability multipliers* form). In the extreme as vulnerability multipliers approaches unity (donor or bottom-up control indicative of a group being close to its carrying capacity), total food consumption rate Q approaches a constant (Ecopath base consumption), so Q/B becomes inversely proportional to B.

Compensatory natural mortality

Compensatory changes in natural mortality rate (M) can be simulated by combining two effects: non-zero *Feeding time adjustment rate* (set on *Ecosim > Input > Group info* form), and either high *EE* from Ecopath or high proportion of *M0* due to predation (unexplained predation > 0). With these settings, especially when vulnerability multipliers of prey to a group are low, decreases in biomass lead to reduced feeding time, which leads to proportional reduction in natural mortality rate.

Attribution This chapter is in part adapted from the unpublished EwE User Guide: Christensen V, C Walters, D Pauly, R Forrest. Ecopath with Ecosim. User Guide. November 2008.

20.

Predator satiation and foraging time

Predator satiation and handling time effects

Ecosim and Ecospace allow you to represent two factors that may limit prey consumption rates per predator (Q/B):

1. foraging time adjustments related to predation risk and/or satiation; and
2. handling time effects.
3. Parameters for both are specified via the Ecosim Group info form.

Satiation and/or choices to forage for short times in order to avoid higher predation risk are represented by setting non-zero values for the *Ecosim > Input > Group info > Feeding time adjustment rate* of a group: larger values of this rate represent more rapid adjustment of foraging time. Non-zero foraging time adjustment rates cause Ecosim/Ecospace to update relative foraging time during each simulation so as to represent predators as trying to maintain Q/B near the Ecopath input base rate. For some organisms (particularly marine mammals) this foraging time adjustment may represent animals always trying to feed to satiation (Q/B from Ecopath the satiation feeding rate) and taking more or less time to reach satiation depending on prey densities (and possibly also facing higher predation risk when foraging times are longer). For other organisms, the Ecopath base Q/B may represent a much lower feeding rate than the animal could achieve under "safe" laboratory conditions, and in this case we view the base Q/B as an evolutionary "target" rate representing results of natural selection for balancing benefits from feeding with predation risk costs of spending more time feeding.

Handling time effects represent the notion that predators have limited time available for foraging and this time can be used up by "handling time" (pursuit/manipulation/ingestion time per prey captured) rather than searching for prey, when prey densities are high. The *Ecosim > Input > Group info > QB$_{max}$/QB$_0$ (for handling time) (>1)* parameter allows you to set ratios of maximum to Ecopath base food consumption rates per individual (or per biomass). These ratios are set to large values (1000) by default, which allows predators to increase their feeding rates without limit as prey densities increase (i.e., not limited by time required to handle each prey). In most scenarios, limitation of prey vulnerability prevents this unreasonable assumption from having noticeable effect. But in scenarios where vulnerable prey densities of at least one type do increase greatly, setting a low value (e.g., 2 or 3) for the predator's maximum/base feeding rate ratio allows you to represent limits on feeding rate associated with time needed to handle each prey. Without such limits, your predictions of increase in predator Q/B, and hence productivity, at low predator density (or high prey density) might be too optimistic and lead you to errors like overestimating sustainable harvest rate for the predator. Also, ignoring handling time effects when one prey type increases greatly can cause an underestimate of the 'buffering' effect that such increases can have on predation rates felt by other prey: if the predator consumes more of the abundant prey, and spends more time handling/resting because of this, predation rates on other prey species should decrease.

Ecosim/Ecospace calculates feeding rates of predators using the "multispecies disc equation", a generalization of Holling's type II functional response model for multiple prey types (see the Holling functional response chapter). Using the maximum/base feeding rate ratio R_j from the *Ecosim > Input > Group info* form along with the Ecopath base food consumption rate per predator, the program calculates a maximum ration and effective handling time per prey biomass eaten (handling time = 1 / (maximum prey biomass eaten per time)). This handling time (Holling's h parameter) is used to calculate the denominator in the disc equation formulation Q_{ij}/B_j , biomass of prey type i consumed per time per unit biomass of predator j, as $Q_{ij}/B_j = a_{ij} V_{ij} / (1 + h_j \sum_k a_{kj} V_{kj})$ where a_{ij} is the rate of effective search by predator j for type i prey, h_j is the predator handling time parameter, V_{ij} is the instantaneous density of prey type i vulnerable to predator j, and the sum in the denominator is over all prey types k taken by the predator. A useful fact about the multispecies disc equation is that D_j, the proportion of time spent feeding (reactive to prey rather than handling), is given by $D_j=1/(1 + h_j \sum_k a_{kj} V_{kj})$. For more information about how the disc equation D_j enters food consumption rate calculations along with other factors that influence feeding, see the Foraging arena chapter. The solution for vulnerable prey densities V_{ij} needed in the disc equation calculation over time involves a numerical procedure that can occasionally cause annoying "chatter" in the Ecosim results when handling times are large (ratio of maximum/base consumption rate small).

A helpful fact about the D_j proportion of time spent feeding in the disc equation formulation is that it can be calculated simply from the user-provided ratio R_j of maximum to ecopath base ration, as just $D_j=R_j/(R_j-1)$. This is used to initialize the D_j at the start of each Ecosim run.

Bioenergetics models for fish most often indicate that feeding rates are low compared to maximum ration; typical ratios of estimated to maximum ration (Hewett-Johnson[1] P parameter) are around 0.3-0.4. These estimates imply R_j (maximum/Ecopath base ration) values of at least 2-4. If you choose to use such realistic values instead of the default 1000, and if this causes Ecosim/Ecospace to show oscillatory behaviour, you need to consider two possibilities:

- The oscillatory behaviour may be a numerical artifact of the procedure used to update D_j; or

- The model's "correct" behaviour for the parameter combinations you have provided is indeed a predator-prey cycle.

If the oscillation has a period of several time steps (months), it is very likely a predator-prey cycle. Persistent predator-prey cycles are commonly predicted by models that include handling time, along with strong top-down control (high vulnerability multipliers v_{ij} of prey to predators). If you think the cycle is unrealistic, you should adjust the prey vulnerabilities multipliers (*Ecosim > Input > Vulnerabilities*) to lower values (toward "bottom up", prey vulnerability control) rather than just setting high R_j values. If you see very short cycles indicating numerical instability in the D_j adjustment procedure (usually happens for fast turnover groups like micro-zooplankton), you should set higher R_j values for the offending groups. This amounts to recognizing that Ecosim may be limited in its ability to represent very fast dynamic changes in groups that turn over very rapidly.

Notes

1. Hewett SW and BL Johnson. 1992. Fish bioenergetics model 2. Univ. of Wisconsin Sea Grant Institute. https://repository.library.noaa.gov/view/noaa/35468

21.

Holling functional response

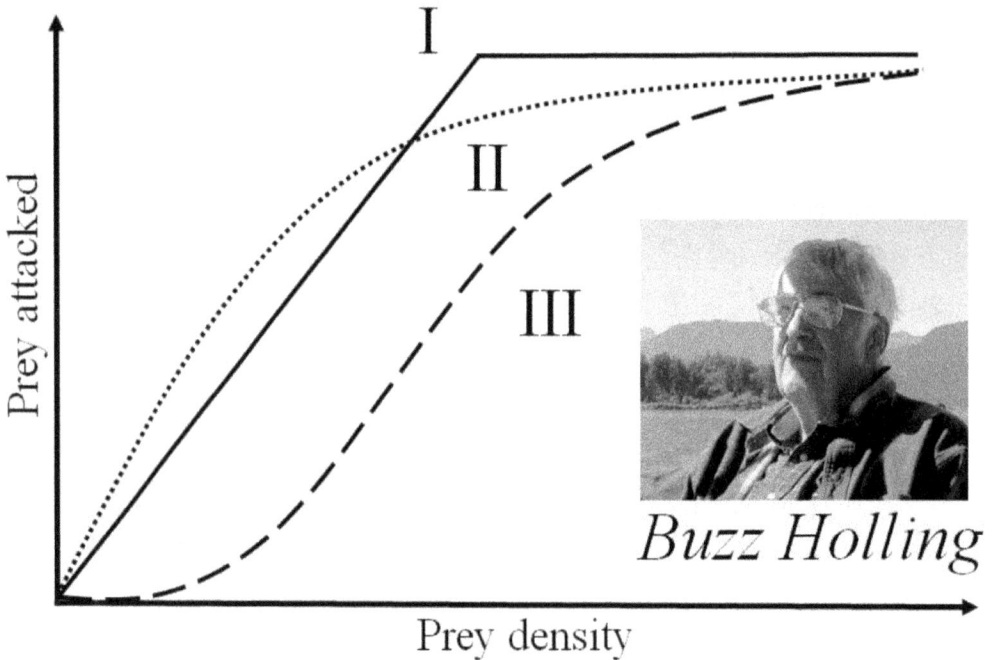

Figure 1. Functional feeding responses as defined by CS Holling.

The Ecosim equations for predicting consumption allow for classic predator-prey functional response types.

- type I if foraging time adjustment is included,
- type II if handling times limit per-capita food-consumption rates, and
- type III if predator rates of search decrease when prey densities are low or prey spend less time foraging when their densities are low

The foraging-arena equations in Ecosim depart from classical functional response predictions in calculating prey densities not as average total biomass in each cell but rather as effective biomass densities in the restricted arenas where foraging typically occurs and where such local densities can be strongly affected by densities of competing predators (ratio-dependence effect). Local depression of available prey biomass can occur whether or not predation affects overall grid-wide prey densities.

Type III switching

Predators are said to "switch" from one prey to another when predator diet proportion of each type changes more rapidly than the relative abundance of that type in the environment. Eating more of something when it becomes abundant does not imply switching, but rather just more frequent encounters with that type. The predator is said to switch if it takes disproportionately more of a prey type as that prey becomes more abundant.

Three mechanisms that can lead to switching patterns in diet composition and prey mortality are represented in Ecosim:

- Apparent switching away from prey that are declining in abundance, due to those prey seeing less intra-specific competition and hence spending less time at risk to predation; this effect occurs for any prey species (and impacts feeding on it by all of its predators) whenever *Ecosim>Input>Group info>Feeding* time adjust. rate is set >0.

- Apparent switching in Ecospace, caused by fitness-sensitive movement; when Ecospace parameters are set to cause increased (and/or directional) movement from cells where "fitness" (per capita food intake minus instantaneous mortality rate) is lower, predators will appear (for the system as a whole) to switch to more abundant prey, and prey that are declining in abundance will see lower predation rates in the cells where they remain concentrated.

- Explicit changes in Ecosim rates of effective search, representing fine-scale behavioural choices by predators to spend more or less foraging time in the arenas where specific prey is concentrated. In this case, the behavioural choice among arenas is predicted from Ideal Free Distribution (IFD) arguments that predators should allocate foraging time so as to minimize time needed to obtain normal food consumption rates.

In the third of these approaches, the Ecosim rate of effective search a_{ij} for predator type j on prey type i is modified at each simulation time step in relation to changes in abundance of all prey types, using a "gravity model" approximation for the IFD allocation of predator foraging time among prey-specific foraging arenas. The equation used for this modification is

$$a_{ij}(t) = K_{ij}a_{ij}B_i(t)^{P_j} / \sum_{i'} a_{i'j}B_{i'}(t)^{P_j} \qquad (1)$$

Here, a_{ij} is the base rate of effective search calculated from Ecopath and vulnerability exchange parameters, K_{ij} is a scaling constant that makes the time-specific $a_{ij}(t)$ equal a_{ij} when all prey biomasses B_i are at the Ecopath base values, and the "switching power parameter" P_j is a user-supplied (empirical, to be estimated from field data or model fitting) power parameter representing how strongly the predator responds to changes in prey availability (switching power parameter on the Group info form). In particular:

- $P_j = 0$, no switching

- $P_j \ll 1$, prey must become very rare before predator j stops searching for them

- $P_j \gg 1$, predator switches violently when any prey increases or decreases.

P_j is limited to the range [0,2]. While it is derived by pretending that predators must allocate time among mutually exclusive foraging arenas for each of their prey types (a typically unrealistic assumption), it can still be used (with $P_j \ll 1$ values) to represent more general ideas about why and how predators switch among prey, e.g., formation and loss of search images for finding them.

There is a tutorial about prey switching included in the Ecosim Group info tutorial (web and pdf versions only)

Attribution The chapter is in part adapted from the unpublished 2008 EwE User Guide and from Walters et al. 2008, *Bulletin of Marine Science*[1] and 2010, which permits authors to use figures, tables, and brief excerpts in scientific and educational works provided that the source is acknowledged and the use is non-commercial.

Notes

1. Walters, C, Martell, SJD, Christensen, V, and Mahmoudi, B. 2008. An Ecosim model for exploring ecosystem management options for the Gulf of Mexico: implications of including multistanza life history models for policy predictions. Bull. Mar. Sci. 83(1): 251-271

22.

Mediation and time forcing

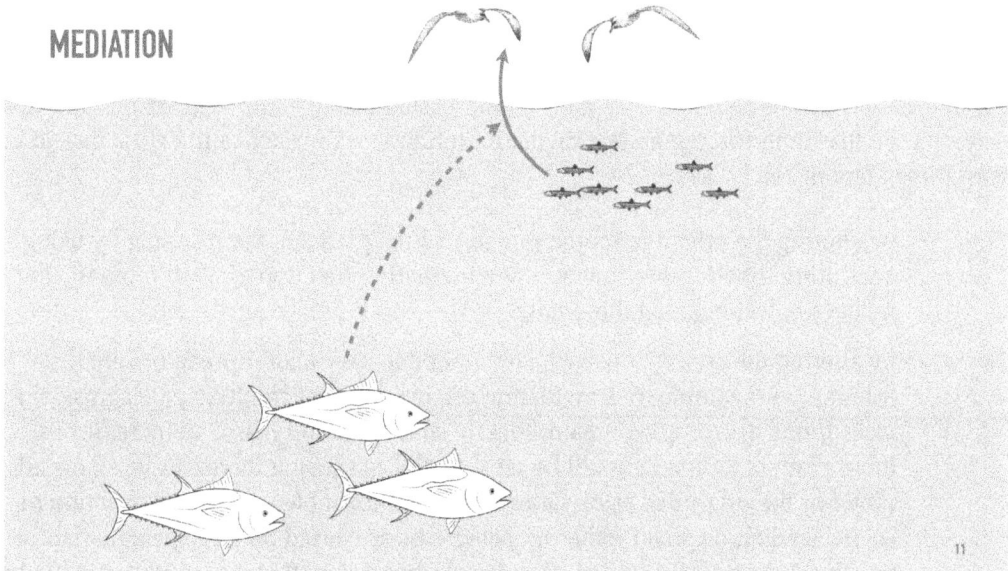

MEDIATION

The basic Ecosim prediction for consumption or flow rate (unit: biomass/time, e.g., t km^{-2} year^{-1}) of type i prey biomass to type j predator is of the functional form

$$\text{flow rate} = a_{ij} V_{ij} B_j \tag{1}$$

where a_{ij} is a "rate of effective search" parameter, V_{ij} is vulnerable prey biomass, and B_j is effective predator abundance (for simple models B_j is just predator biomass; for multi-stanza groups it is the sum over ages in that group of numbers at age times body weight to the 2/3 power, an index of per-predator search rate). If vulnerable prey were randomly distributed over the modelled spatial area, and V, B_j were expressed as abundances per unit area, then a_{ij} would be interpretable as a volume or area swept per unit predator abundance (per B_j) per unit time, corrected for the proportion of time actually spent searching for food (foraging time and handling time adjustments reduce a_{ij} from its theoretical maximum value for a predator that searched all the time for food).

To understand how effects of habitat changes as represented through time forcing functions, and mediation effects as expressed through mediation functions of abundances of other organisms, are likely to affect trophic flow rates, we need to be a bit more careful about the a_{ij} parameter. In particular, we need to recognize that for most trophic interactions, predators search for prey only over restricted spatial foraging arenas, and hence V_{ij} is distributed only over such areas rather than at random over the whole system.

Suppose the (practically unmeasurable) restricted area where foraging by j on prey i takes place is A_{ij} per unit total model area. Suppose that while in this area, each unit of predator abundance (per B_j) searches an effective area a_{ij}^* for prey. On average, each such area searched should result in capture of V_{ij}/A_{ij} prey, since this ratio is prey density in the arena area. In other words, the flow rate could be modelled more precisely (if we could measure A_{ij}) as

$$\text{flow rate} = a_{ij}^* A_{ij} V_{ij} B_j \tag{2}$$

i.e., the basic Ecosim equation's a_{ij} can be interpreted as $a_{ij} = a_{ij}^* / A_{ij}$. Expressed this way, we see that time forcing and/or mediation effects can influence the flow rate in at least three quite distinct ways:

- by altering the effective search rate a_{ij}^* of the predator, for example by using a turbidity time forcing function or a mediation function of algal biomass that reduces a_{ij}^* at high algal biomass.

- by altering the area A_{ij} over which vulnerable prey and/or predators are distributed, for example by a mediation effect where macrophyte or seagrass biomass limits the foraging area usable by small predatory fish, so increases in those plant biomasses should be represented as causing increases in A_{ij} for all prey i of the small fish as predator j. Another example would be restriction of A_{ij} for feeding on small fishes by pelagic birds caused by large pelagic fishes, which drive small fishes nearer to the surface where they are more available to the birds.

- by altering the vulnerability exchange rates v_{ij} that determine (along with a_{ij}^* / A_{ij}) V_{ij} from total prey biomass B_i (the basic equation for V from B is $V_{ij} = v_{ij} B_i / (v_{ij} + v_{ij}{}' + a_{ij}^*/A_{ij} B_j)$). For example, if small fish respond to increased large plant biomass by occupying a larger area, the mixing rate (v_{ij}) of planktonic food organisms into that larger area will increase as well.

It is possible to apply multiple time forcing and/or mediation functions to each trophic flow (i,j) rate prediction, and to specify whether each function multiplies a_{ij}^*, A_{ij}, and/or v_{ij}. Using these forms, one can choose the parameter that is multiplied by each forcing or mediation function, i.e. one of the following choices:

- Multiply overall predator rate of effective search ($a_{i,j}$), for example to represent time-varying turbidity changes that affect predator search efficiency or mediation effects of algal biomass on search efficiency.

- Multiply vulnerability exchange rate (v_{ij}), for example to represent increased movement rates of prey into vulnerable behavioural state at times when water mixing rates are higher.

- Multiply area of foraging arenas (divide A_{ij} by multiplier), for example to represent increase in habitat area available for juvenile fish refuges.

- Multiply area (divide A_{ij}) and also multiply v_{ij}, for example to represent increase in safe foraging habitat available to a predator that feeds on prey that become available in foraging arenas through passive drift/mixing processes such that increasing area used by predator results in higher proportion of total prey population being available in foraging areas at any moment.

Forcing and trophic mediation functions in Ecosim can be set up at the *Ecosim > Input > Forcing function; and Ecosim) > Input > Mediation* forms.

23.

Dynamic instability and multiple stable states

Prey fish   Piscivore

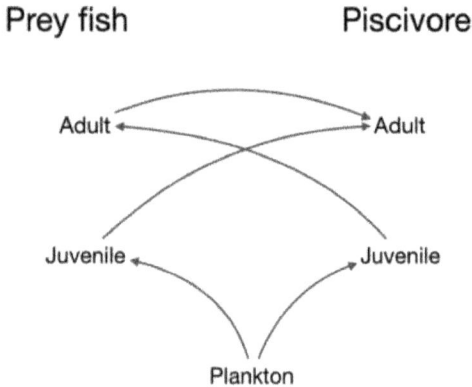

Figure 1. Complex ontogenetic feeding relationship impacting predator-prey balance. A predator may cultivate the environment for its juveniles by feeding on a prey that competes and perhaps prey on its own juveniles. If the predator abundance is reduced, e.g., by fishing, the prey abundance may increase, potentially making it difficult for the predator to rebound. In models we often see alternate stable states occurring when a prey species feed on the young stages of its predators.

Dynamic instability in Ecosim and Ecospace

We commonly see several types of dynamic instability following small perturbations in fishing mortality rates (to get away from initial Ecopath equilibrium):

- Predator-prey cycles and related multi-trophic level patterns;

- System simplification (loss of biomass pools due to competition/predation effects);

- Stock-recruitment instabilities (cyclic or erratic changes in recruitment and stock size for split pool groups);

- Numerical "chatter" in time solutions (mainly in Ecospace).

Such patterns are not particularly common in fisheries time series, so unless you have data to support a cyclic prediction, it's probably reasonable to adjust the model parameters to get rid of it.

Predator-prey and simplification effects can usually be eliminated by reducing the predation vulnerability multipliers *(Ecosim > Input > Vulnerability multipliers* form, set values to 4 or less).

We know of at least four common mechanisms that can decrease the vulnerability multipliers so as to create stabilizing and the appearance of "ratio-dependent" or "bottom-up" control of consumption rates:

- *Risk-sensitive prey behaviours*: Prey may spend only a small proportion of their time in foraging arenas where they are subject to predation risk, otherwise taking refuge in schools, deep water, littoral refuge sites, etc.

- *Risk-sensitive predator behaviours* (the "*three to tango*" argument): Especially if the predator is a small fish, it may severely restrict its own range relative to the range occupied by the prey, so that only a small proportion of the prey move or are mixed into the habitats used by it per unit time. In other words, its predators may drive it to behave in ways that make its own prey less vulnerable to it.

- *Size-dependent graduation effects*: Typically a prey pool represents an aggregate of different prey sizes, and a predator can take only some limited range of sizes, limited vulnerability can represent a process of prey graduation into and out of the vulnerable size range due to growth. Size effects may of course be associated with distribution (predator-prey spatial overlap) shifts as well.

- *Passive, differential spatial depletion effects*: Even if neither prey or predator shows active behaviours that create foraging arena patches, any physical or behavioural processes that create spatial variation in encounters between predator and prey will lead to local depletion of prey in high-risk areas and concentrations of prey in partial predation "refuges" represented by low-risk areas. "Flow" between low and high-risk areas (v_{ij}) is then created by any processes that move organisms.

These mechanisms are so ubiquitous that any reader with aquatic natural history experience might wonder why anyone would ever assume a mass-action, random encounter model (high vulnerability multipliers (e.g., 100) in the *Ecosim > Input > Vulnerability multipliers* form) in the first place.

Methods for dealing with stock-recruitment instability are discussed in the help section on using Ecosim to study compensation. Generally, the simplest solutions are to check (and reduce if needed) cannibalism rates, set higher foraging time adjustment rates (*Ecosim > Input > Group info* form) for juvenile pools and reduce vulnerabilities of prey to juvenile fishes (*Ecosim > Input > Vulnerability multipliers* form).

Numerical instabilities (chatter, oscillations of growing amplitude) occur mainly in Ecospace. They are avoided in Ecosim by only doing time dynamic integration of change for pools that can change relatively slowly. In Ecospace, the only remedy for chatter is to reduce the prediction time step (from 12/year default value, sometimes very low values such as 0.05 year are required for stability). In extreme cases, it might be necessary to "fool" Ecosim and Ecospace by implicitly moving to a shorter time step for all dynamics, which you can do by dividing every Ecopath input time rate (*P/B*, *Q/B*) with the same factor.

Multiple stable states

With careful parameter choices, Ecosim can also represent Holling's resilience concept of multiple stable states. In particular, so-called "cultivation-depensation"[1] or "trophic triangle" effects can lead to stable states dominated by large or small species, with the initial Ecosim equilibrium being an unstable or saddle-point between these states, (see Figure 1). For example, suppose a dominant large species like cod is fished down, and that one or more of its prey species then increase. If the prey species, then consume or compete with juveniles of the cod, cod juvenile survival rates may be reduced enough to cause the cod to continue to collapse, leading to a stable equilibrium dominated by the smaller species. As a historical note, the very first Ecosim model was developed by Alida Bundy for Manila Bay in the Philippines. That model exhibited two stable states for two competing groups of small Leiognathid fishes, with small perturbations from the initial Ecosim state lead to dominance by one or other of the two groups.

Attribution This chapter is in part adapted from the unpublished EwE User Guide: Christensen V, C Walters, D Pauly, R Forrest. Ecopath with Ecosim. User Guide. November 2008.

Notes

1. Walters C and Kitchell JF. 2001. Cultivation/depensation effects on juvenile survival and recruitment: implications for the theory of fishing. Can. J. Fish. Aquat. Sci. 58: 39–50. https://doi.org/10.1139/f00-160

24.

Hatchery production

Multi-stanza populations can be designated as hatchery populations, and hatchery production can be varied over time using time forcing functions. To turn off natural reproduction and replace it with a time series of hatchery stocking rates, open the *Ecopath > Input > Edit multi-stanza* groups form for a population, and enter a nonzero value for the hatchery forcing function number. Forcing functions can be sketched using the *Ecosim > Input > Forcing function* form or can be imported with time series (*Ecosim > Input > Time series*). Check the Forcing function form for the number allocated to your hatchery stocking time series. Forcing functions to represent historical changes in stocking rates can be entered via the same CSV files as used to set up historical fishing and model fitting scenarios. Enter stocking rates as values relative to the stocking rate of 1.0 assumed for the Ecopath base year.

Then at each simulation time step, the base recruitment for the population (calculated from Ecopath input parameters) will be multiplied by the current time value for the designated forcing function. A forcing function value of 1.0 corresponds to the stocking rate that would result in the Ecopath base abundance (biomass) entered.

If it is desired to simulate stocking of older fish at some age like 18 months, the first stanza for the population should be set to have this duration, the mortality rate (Z or P/B) for the stanza should be set to 0.001, and the diet for the stanza should be set to 1.0 imported (i.e., do not have fish in the stanza feeding in the modelled ecosystem).

Attribution This chapter is in part adapted from the unpublished EwE User Guide: Christensen V, C Walters, D Pauly, R Forrest. Ecopath with Ecosim. User Guide. November 2008.

25.

Non-additive mortality rates

Consider how total mortality (Z, year^{-1}) commonly is estimated in fisheries or ecosystem models,

$$Z = F + M0 + M2 \qquad (1)$$

where F (year^{-1}) is fishing mortality, $M2$ (year^{-1}) predation mortality, and $M0$ (year^{-1}) "other mortality", i.e. the total mortality rates not due to fisheries (included in the model) and predation (as included in the model). The question then is, what will happen if the predation mortality is decreased, e.g., due to targeted reduction in predator populations?

Your immediate answer to that question could well be that if predation is reduced then the total mortality would be reduced as well. That would result in more of the species of interest surviving to recruitment and beyond. Indeed, ecosystem models typically assume additive effects of predation and other natural mortality rates in prediction of net production for small forage fishes in particular, resulting in prediction of substantial increase in forage fish production when predator abundances are reduced. But what if vulnerability to predation is affected by stress factors (e.g., hunger and parasite loads) that would result in higher mortality rates of vulnerable forage fish individuals even if predators were removed? In that case there may in fact be little decrease in forage fish natural mortality rates and hence little or no increase in net production rates.

Ecosystem models that account for trophic interaction effects on prey (e.g. forage fish) mortality rates very typically represent mortality rates as a sum of independent or additive component rates, with a rate component for each predator type (species, size) that is determined by prey and predator abundances and with some constant non-predation or "other" mortality component. Such formulations ignore that prey individuals taken by predators may be predominantly those vulnerable to predation because of behavioural or physiological stress factors (e.g. hunger, parasite or disease load, physiological effects of aging and/or spawning) that would kill a proportion of the vulnerable individuals even if predators did not take them. The existence of such stress factors, and concentration of both predation and other mortality on individuals made vulnerable by them, implies that mortality rate components should not be treated as additive. Parasites and pathogens in particular may exert strong regulatory effects on trophic interactions in general and predation mortality rates in particular[1][2][3].

Another key stress factor leading to increased predation vulnerability may be contaminant loading[4]. Explicit representation of how such stress factors can lead to increased mortality could lead to more realistic and useful models in cases where such effects are now represented by ad hoc approaches, e.g. to starvation rates and quadratic mortality terms representing increasing mortality rate at higher abundances.

The assumption of additive predation mortality rate impacts on forage fish in particular results in predictions of substantial increase in surplus production of these small fish when piscivore abundances are reduced through fishing or appropriation of forage fish production by fisheries, since a high proportion of the forage fish natural mortality rate is typically estimated to be due to predation (see, e.g.,[5][6]. This increase in predicted net production (e.g.,[7][8]) occurs in both simple biomass dynamics models like Ecosim[9] and in more detailed size spectrum models[10][11][12], and is a key reason for predicted increases in yield under balanced harvesting policies[13].

Non-additivity of mortality rates can be represented very crudely in Ecosim as foraging arena limitations on predation rates[14][15]. Surplus production rate predictions for forage fish under such circumstances can result in much weaker predicted responses of production rate to decreases in predator abundances than are now obtained with models like Ecosim or size spectrum models. We[16] developed one way to represent non-additivity hypotheses in Ecosim, and used an empirical example involving possible non-additive effects of pinniped predation on juvenile Chinook and coho salmon in the Georgia Strait, British Columbia to demonstrate how uncertain predictions of impact of changing predator abundance can be if measured predation rates are in fact limited by stress factors that would cause high mortality rates even if predator abundances were much reduced.

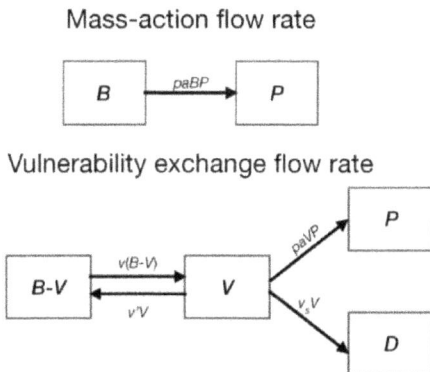

Figure 1. Alternative approaches to prediction of mass flow rate from any one prey biomass component B and predator component P. See text for explanation and Table 1 for parameter definitions.

Vulnerability exchange model

There are at least two alternative approaches to prediction of biomass flow rate along any ecosystem model link between a prey biomass component (species, size) B and a predator biomass component (species, size) P (Figure 1). In the mass-action or spatially mixed approach used in existing size spectrum models and other approaches like the Essington and Munch[17] equilibrium-perturbation model, flow to the predator (consumption rate as mass per time) is assumed proportional to prey biomass and predator biomass, with proportionality constant pa, where a is the predator rate of effective search and p is the proportion of time spent searching by the predator.

Table 1. Parameter definitions. In the Units column, t = time

Parameter	Definition	Units
a	Predator rate of effective search	area $P^{-1} t^{-1}$
B	Prey biomass	mass/area
D	Detritus biomass	mass/area
h	Time lost from predator searching per unit prey biomass eaten	t/mass
Mx	Instantaneous mortality rate due to factor x	t^{-1}
$M0$	Base instantaneous total mortality rate not due to predation (other mortality)	t^{-1}
P	Predator biomass	mass/area
p	Proportion of time spent searching by predators	dimensionless
r_{max}	Forage fish maximum production rate at high B	mass/t
v	Rate at which prey become vulnerable to predation	t^{-1}
v'	Rate at which prey recover or move to safe areas	t^{-1}
v_s	Rate at which vulnerable prey die due to stress factor(s)	t^{-1}

If the predator is assumed to have a type II functional response where handling time may limit its feeding rate[18], p is assumed to vary with the abundances B_i of all prey types taken by the predator type, as,

$$p = 1/(1 + h \sum_i a_i B_i) \qquad (2)$$

where h is handling time lost from searching per unit of prey biomass consumed.

In the foraging arena or vulnerability exchange model, prey are assumed to move or flow between invulnerable and vulnerable behavioral states at instantaneous rates v and v', with predation and stress-related loss rates predicted to occur only from the vulnerable biomass component V. The original Ecosim models assume this structure for all trophic links, with the stress-related mortality rate vs set to 0, and with the exchange process assumed to be created either by restricted predator activity (mixing of prey into and out of restricted spatial areas (foraging arenas) where predators forage) or by restricted prey activity where prey become vulnerable through foraging activities that force them to leave invulnerable refuge habitats[19][20].

In Figure 1, we have added a direct, stress-related mortality component vsV, to represent the possibility that the flow rate $v(B-V)$ into vulnerable states represents prey and predator foraging restrictions and possibly also actions of stress agents, and flow rates from V back to B-V also include a loss rate vs representing mortality caused by those stress agents (like growing parasite loads and contraction of diseases). When v arises at least partly from such stressors, v' represents recovery rate due to processes like parasite shedding and recovery from disease.

Whether or not there is a direct stress-related mortality rate component, dynamics of vulnerable biomass V can be represented by the continuous rate model

$$dV / dt = v(B - V) - (v' + paP + v_s)V \tag{3}$$

If the vulnerability exchange process is relatively rapid compared to rates of change in B, i.e. if the instantaneous loss rate $v'+paP+v_s$ is large, then predicted V will vary with B so as to remain near the moving equilibrium given by setting $dV/dt = 0$ and solving eq. (2) for V, i.e. by

$$V = vB / (v + v' + paP + v_s) \tag{4}$$

That is, V will be proportional to B and inversely proportional to v plus the total instantaneous loss rate.

Note that Eq. 4 predicts a maximum total biomass flow rate to predation and stress mortality $(paP+v_s)V$ to have an upper bound vB, i.e. a maximum rate set by how rapidly the prey become vulnerable to P due to behavior and stress. Further, it predicts that the flow rate to the predator, $paPV$, should be inversely related to the stress mortality rate vs, and the direct flow rate to stress mortality (v_sV) should be inversely related to predator abundance P, i.e. the two mortality rate components should not be independent of one another.
  This trade-off between mortality components can be very severe if both paP and v_s are large (Figure 2).

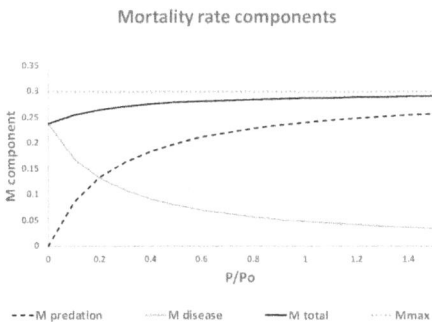

Figure 2. Effect of varying predator biomass P on the total mortality rate M = *flow/total biomass B* due to predation and stress, for a case where both the instantaneous predation rate paP and stress rate v_s are large. In this example, total mortality rate remains close to the vulnerability exchange rate v, even when predator biomass P is zero. See Table 1 for parameter definitions.

If we compare instantaneous mortality rates MP due to predator P for the mass-action and vulnerability exchange formulations as in Figure 2, where MP = (mass eaten per time) / (total prey biomass), we obtain very different predictions about how MP should vary with predator abundance:

Mass-action model:

$$M_p = paP \qquad (5a)$$

Vulnerability exchange model:

$$M_p = \frac{paPV}{B} = \frac{paPv}{v + v' + paP + v_s} \qquad (5b)$$

That is, we predict mortality rate to be additive to other mortality rate components and proportional to predator abundance only for cases where the prey are fully vulnerable to predation at all times and are not subject to stress-related mortality, i.e. for cases where v is very large and $v_s = 0$. For the vulnerability exchange case, we predict the total instantaneous mortality rate for the B–P flow link to vary as

$$M_{total} = \frac{(paP + v_s)V}{B} = \frac{(paP + v_s)v}{v + v' + paP + v_s} \qquad (6)$$

for which the mortality flow components $paPV$ and v_sV are very obviously not additive because of their joint, interacting effect on V.

Predator abundances and forage fish surplus production

To examine how forage fish surplus production rates should vary when M_{total} from Eq. 6 is used to predict the joint effects of predation and stress, suppose we now consider a simple case where there is only a single aggregate predator abundance P and/or all species-size components of the total predation rate are assumed to covary so as to generate a single overall rate that is proportional to the sum of the predator biomass components. Suppose further that we assume the dynamics of B to be dominated by a production-recruitment component that can be adequately described by a Beverton-Holt recruitment function, minus a non-predation mortality rate $M0\,B$ minus the total rate $M_{total}B$ predicted by Eq. 7, below. Here, $M0$ is the direct non-predation mortality rate as M_{total} include indirect effects of disease/stress contributions. Under these assumptions, the dynamics of B are given by

$$dB / dt = (r_{max}B) / (B_h + B) - (M0 + M_{total}B) \qquad (7)$$

where r_{max} is maximum production rate (mass/time), B_h is the forage fish biomass needed to achieve half of this maximum rate, and M_{total} is given for varying P by Eq. 6. Note in Eq. 7 that dB/dt represents the surplus production rate of the forage fish population.

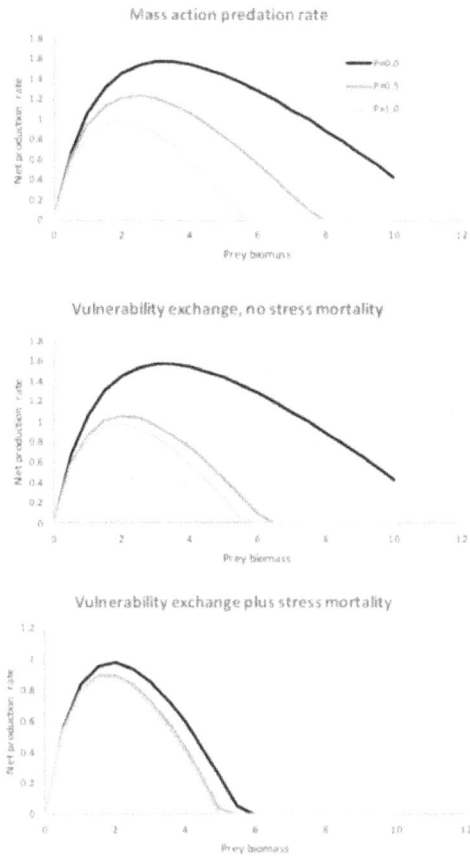

Mass action predation rate

Vulnerability exchange, no stress mortality

Vulnerability exchange plus stress mortality

Figure 3. Predicted patterns of variation in the relationship between prey (e.g., forage fish) surplus production rate (*dB/dt* of Eq. 7) and biomass for different approaches to prediction of predation rates and for varying predator abundances (*P*, black lines are for high *P*, light gray for *P* = 0.0).

Very different patterns of variation in the surplus production vs biomass relationship are predicted by Eq. 7 depending on how predation and stress mortality is represented (Figure 3). For the mass action case ($M_{total} = paP$) and for vulnerability exchange dynamics without stress factor removal from the vulnerable biomass V, Eq. 7 predicts substantial increase in surplus production when P is reduced. But when $Mtotal$ includes a high mortality rate due to stress when P is low, as in Figure 2, there is almost no response of the predicted surplus production relationship due to reduction in predator abundance.

Note that the patterns predicted in Figure 3 depend importantly on the assumption that foraging time proportion p is stable, i.e. that either there are no handling time effects or that B is a relatively small proportion of the total prey abundance that contributes to predator handling time. When search time does increase substantially at low B, there can be severe depensatory effects that cause reduced or even negative surplus production rates when prey biomass is low.

Simple approximation for non-additive mortality effects

It would be a complex conceptual and programming task to fully represent non-additive mortality patterns like the example in Figure 1 in Ecosim models even for simple stress mortality rate assumptions like $v_s V$ with constant v_s, because of issues about partitioning Ecopath base unexplained/ other mortality' ($M0$), and whether to assume different stress-related vulnerability and mortality patterns for each of many predator-prey trophic links in typical Ecosim models. However, the same basic effect as shown in the third panel of Figure 3 can be obtained simply by first calculating the Ecosim predicted predation mortality rate (summed over predator types) at each time step, then adding a component to $M0$ so as to prevent the total "apparent" predation mortality rate from decreasing to less than a user-defined minimum proportion of the Ecopath base predation mortality rate. This simple convention allows exploration of alternative hypotheses about how much $M0$ would increase in scenarios where total predator abundance is reduced, using a single parameter (proportion of base predation mortality rate when predation rate is 0.0) to represent "hidden" non-additive (stress factor) effects.

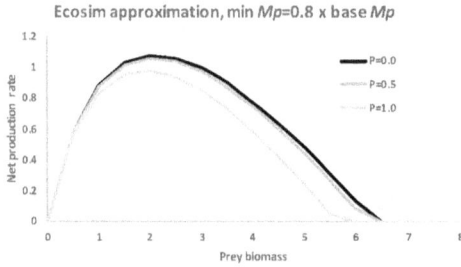

Figure 4. Effect of including a minimum mortality rate component representing non-additive stress mortality on Ecosim predictions of the surplus production vs biomass relationship. Compare this prediction to those in the bottom two panels of Figure 3 – the simple mortality rate constraint causes the Ecosim prediction to be close to that obtained by explicitly modeling non-additive stress mortality.

Effects of such a minimum mortality constraint on the predicted surplus production-biomass relationship for the same parameter values used in Figure 3 is shown in Figure 4. Setting the minimum "hidden" predation rate to 0.8 times the Ecopath base rate causes the predicted surplus production pattern to look almost exactly like the full non-additive mortality pattern in the bottom panel of Figure 3, i.e. to predict very little increase in surplus production when predator abundance is low.

As the ratio of minimum to Ecopath base apparent predation mortality rate is reduced (from 0.8 to lower values for the Figure 4 example), the predicted surplus production pattern shifts to equal the pattern predicted in the middle panel of Figure 3 as the ratio approaches zero.

We caution against using this single-parameter approach just to generate multispecies scenarios where trophic interaction effects are omitted entirely (by setting the additive proportion very low), just to conform with single species modeling theory and experience.

Discussion

It is an old idea in ecology going back at least to Errington[21] that predators may take mainly weak, sick, and old animals, and that it therefore may be perilous to assume that predator control will increase productivity of valuable prey species. In terms of current terminology about top-down (predation) versus bottom-up (prey productivity) control of trophic interactions, low values of the vulnerability exchange rate parameter v in the model presented above imply stronger bottom-up control of predator abundances, while high v values imply at least the possibility of strong top-down control but with the caveat that high vs values may invalidate predictions based just on v and on predator abundance. Our models warn not to expect correct or reliable assessments of relative importance of top-down vs bottom-up effects from correlative studies of abundance variation over time (e.g.[22 23 24] because of possible interactive effects of predation and bottom-up "habitat" factors like temperature, as we estimate to be possible for the salmon example.

In evaluation of evidence about the relative importance of top-down effects, we should focus mainly on cases where there have been deliberate manipulations of top-down effects (e.g.,[25 26] that would reveal existence of vs effects if such effects are indeed present. This warning holds as well for those rare cases where we have direct estimates of variation in natural mortality rates (M) over time from data such as survey relative abundances at age or tagging as in the salmon case; good correlations of M with predator abundance do not imply top-down control when stress factors have changed over time in patterns correlated with predator abundance.

Modern molecular techniques offer considerable promise to screen for gene activation patterns (gene expression profiles) indicative of stress, and hence to directly measure v and V, i.e. whether the prey taken by predators are indeed mainly those that are stressed particularly by diseases (e.g.[27 28 29]. But unfortunately, such techniques do not provide direct measures of the virulence of the stress factors, i.e. of the direct stress mortality and recovery rate parameters v_s and v'; predation impacts may still be essentially additive components of total M if vs is low.

Another way to examine the credibility of hypotheses about non-additive predation impacts is to compare estimates of predator rates of effective search (a) implied by high vs-low V models with direct estimates of rates of search based on predator characteristics. Such direct estimates can be obtained from basic information on predator movement speeds, reactive distances to prey, and proportions of time spent foraging, combined with information on the effective area or volume over which the search is distributed. For the juvenile salmon-seal example in Walters and Christensen[30], such calculations suggest much lower a parameter values for seals than would be necessary to explain the data under high vs assumptions.

As data sets accumulate with age-specific survey estimates of abundance (from which temporal variation in total natural mortality rate M can be estimated), we will also be able to directly compare observed changes in M with predictions from additive predation models (and direct estimates of search rates). Seeing lower slopes in M vs predator abundance plots than predicted under additive predation would be evidence of high stress-dependent mortality rates (vs), as in the Figure 2 example.

Continued climate change will quite possibility result in substantial changes in trophic interaction patterns[31] through "hidden" effects due to temperature-related changes in v and vs, (e.g., increases in disease expression or physiological impact). But such changes may be "masked" when stressed individuals are rapidly removed by predators, so as to only exert increasing effects when various factors, like fishing, lead to predator abundance declines. This means that climate change is quite likely to produce some very nasty surprises that we will not anticipate through ecosystem models built around simplistic assumptions about additivity of mortality components, nor can we be confident that simpler models based on statistical or correlative historical data will somehow give better predictions.

Attribution Based on Walters and Christensen[32]. Excerpts and figures used with permission from Elsevier, License Numbers 5663220213981 and 5663220407849.

Notes

1. Hatcher, M.J., Dick, J.T., Dunn, A.M., 2012. Diverse effects of parasites in ecosystems: linking interdependent processes. Frontiers in Ecology and the Environment 10, 186–194. https://doi.org/10.1890/110016

2. Krkošek, M., 2017. Population biology of infectious diseases shared by wild and farmed fish1. Can J Fish Aquat Sci 74, 620–628. https://doi.org/10.1139/cjfas-2016-0379

3. Sures, B., Nachev, M., Pahl, M., Grabner, D., Selbach, C., 2017. Parasites as drivers of key processes in aquatic ecosystems: Facts and future directions. Exp. Parasitol. 180, 141–147. https://doi.org/10.1016/j.exppara.2017.03.011

4. Gray, R., Fulton, E., Little, R., Scott, R., 2006. Ecosystem model specification with an agent based framework. Technical report CSIRO. Marine and Atmospheric Research. North West Shelf Joint Environmental Management Study; no. 1–139.

5. Engelhard, G.H., Peck, M.A., Rindorf, A., C Smout, S., van Deurs, M., Raab, K., Andersen, K.H., Garthe, S., Lauerburg, R.A.M., Scott, F., Brunel, T., Aarts, G., van Kooten, T., Dickey-Collas, M., 2014. Forage fish, their fisheries, and their predators: who drives whom? ICES JMS 71, 90–104. https://doi.org/10.1093/icesjms/fst087

6. Koehn, L.E., Essington, T.E., Marshall, K.N., Kaplan, I.C., Sydeman, W.J., Szoboszlai, A.I., Thayer, J.A., 2016. Developing a high taxonomic resolution food web model to assess the functional role of forage fish in the California Current ecosystem. Ecol Model 335, 87–100. https://doi.org/10.1016/j.ecolmodel.2016.05.010

7. Walters, C.J., Christensen, V., Martell, S.J., Kitchell, J.F., 2005. Possible ecosystem impacts of applying MSY policies from single-species assessment. ICES JMS 62, 558–568. https://doi.org/10.1016/j.icesjms.2004.12.005

8. Szuwalski, C.S., Burgess, M.G., Costello, C., Gaines, S.D., 2017. High fishery catches through trophic cascades in China. Proc. Natl. Acad. Sci. U.S.A. 114, 717–721. https://doi.org/10.1073/pnas.1612722114

9. Walters, C., Christensen, V., Pauly, D., 1997. Structuring dynamic models of exploited ecosystems from trophic mass-balance assessments. Rev Fish Biol Fisheries 7, 139–172. https://doi.org/10.1023%2fa%3a1018479526149

10. Scott, F., Blanchard, J.L., Andersen, K.H., 2014. mizer: an R package for multispecies, trait-based and community size spectrum ecological modelling. Methods in Ecology and Evolution 5, 1121–1125. https://doi.org/10.1111/2041-210X.12256

11. Jacobsen, N.S., Essington, T.E., Andersen, K.H., 2015. Comparing model predictions for ecosystem-based management1. Can J Fish Aquat Sci 73, 666–676. https://doi.org/10.1139/cjfas-2014-0561

12. Jacobsen, N.S., Thorson, J.T., Essington, T.E., 2019. Detecting mortality variation to enhance forage fish population assessments. ICES JMS 76, 124–135. https://doi.org/10.1093/icesjms/fsy160

13. Garcia, S.M., J Kolding, J Rice, Rochet, M.-J., Zhou, S., Arimoto, T., Beyer, J.E., Borges, L., Bundy, A., Dunn, D., Fulton, E.A., Hall, M., M Heino, Law, R., M Makino, Rijnsdorp, A.D., Simard, F., Smith, A.D.M., 2012. Reconsidering the consequences of selective fisheries. Science 335, 1045–1047. http://dx.doi.org/10.1126/science.1214594

14. Walters et al. 1997, https://doi.org/10.1016/j.ecolmodel.2019.108776 *op. cit.*

15. Ahrens, R.N.M., Walters, C.J., Christensen, V., 2012. Foraging arena theory. Fish Fish. 13, 41–59. https://doi.org/10.1111/j.1467-2979.2011.00432.x

16. Walters C, Christensen V. 2019. Effect of non-additivity in mortality rates on predictions of potential yield of forage fishes. Ecological Modelling, 410: #108776. https://doi.org/10.1016/j.ecolmodel.2019.108776

17. Essington, T.E., Munch, S.B., 2014. Trade-offs between supportive and provisioning ecosystem services of forage species in marine food webs. Ecol Model 24, 1543–1557. https://doi.org/10.1890/13-1403.1

18. Holling, C.S., 1959. The components of predation as revealed by a study of small mammal predation of the European pine sawfly 91, 293–320. https://doi.org/10.4039/Ent91293-5

19. Walters et al., 1997. https://doi.org/10.1016/j.ecolmodel.2019.108776 *op. cit.*

20. Walters, C.J., Juanes, F., 1993. Recruitment limitation as a consequence of natural selection for use of restricted feeding habitats and predation risk taking by juvenile fishes. Can J Fish Aquat Sci 50, 2058–2070. https://doi.org/10.1139/f93-22

21. Errington, P.L., 1946. Predation and Vertebrate Populations (Concluded). The Quarterly Review of Biology 21, 221–245. https://doi.org/10.1086/395315

22. Boyce, D.G., Frank, K.T., Worm, B., Leggett, W.C., 2015. Spatial patterns and predictors of trophic control in marine ecosystems. Ecol Lett 18, 1001–1011. https://doi.org/10.1111/ele.12481

23. Lynam, C.P., Llope, M., Möllmann, C., Helaouët, P., Bayliss-Brown, G.A., Stenseth, N.C., 2017. Interaction between top-down and bottom-up control in marine food webs. Proc. Natl. Acad. Sci. U.S.A. 114, 1952–1957. https://doi.org/10.1073/pnas.1621037114

24. Ye, Y., Carocci, F., 2019. Control mechanisms and ecosystem-based management of fishery production. Fish and Fisheries 20, 15–24. https://doi.org/10.1111/faf.12321

25. Borer, E.T., Halpern, B.S., Seabloom, E.W., 2006. Asymmetry in community regulation: effects of predators and productivity. Ecology 87, 2813–2820. https://doi.org/10.1890/0012-9658(2006)87[2813:aicreo]2.0.co;2

26. McClanahan, T.R., Muthiga, N.A., Coleman, R.A., 2011. Testing for top-down control: can post-disturbance fisheries closures reverse algal dominance? Aquatic Conservation: Marine and Freshwater Ecosystems 21, 658–675. https://doi.org/10.1002/aqc.1225

27. Jeffries, K.M., Hinch, S.G., Gale, M.K., Clark, T.D., Lotto, A.G., Casselman, M.T., Li, S., Rechisky, E.L., Porter, A.D., Welch, D.W., Miller, K.M., 2014. Immune response genes and pathogen presence predict migration survival in wild salmon smolts. Mol Ecol 23, 5803–5815. https://doi.org/10.1111/mec.12980

28. Miller, K.M., Teffer, A., Tucker, S., Li, S., Schulze, A.D., Trudel, M., Juanes, F., Tabata, A., Kaukinen, K.H., Ginther, N.G., Ming, T.J., Cooke, S.J., Hipfner, J.M., Patterson, D.A., Hinch, S.G., 2014. Infectious disease, shifting climates, and opportunistic predators: cumu-

lative factors potentially impacting wild salmon declines. Evol Appl 7, 812–855. https://doi.org/10.1111/eva.12164

29. Tucker, S., Li, S., Kaukinen, K.H., Patterson, D.A., Miller, K.M., 2018. Distinct seasonal infectious agent profiles in life-history variants of juvenile Fraser River Chinook salmon: An application of high-throughput genomic screening. PLoS ONE 13, e0195472. https://doi.org/10.1371/journal.pone.0195472

30. Walters and Christensen. 2019. https://doi.org/10.1016/j.ecolmodel.2019.108776 *op. cit.*

31. Lynam et al., 2017. https://doi.org/10.1073/pnas.1621037114 *op. cit.*

32. Walters and Christensen, 2019, https://doi.org/10.1016/j.ecolmodel.2019.108776 *op. cit.*

26.

Shared foraging arenas

The basic Ecosim formulation for predation interactions assumes that each non-zero consumption of a prey type i by a predator type j takes place in a foraging arena unique to that interaction[1][2][3]. The rationale for this assumption is that each arena is defined by the combined behaviors of both the prey and the predator, and possibly also by selection of particular prey sub-types, (e.g., sizes), such that multiple predators can feed on the same prey type in different ways (at different depths, times of day, spatial microhabitats) without competing directly for prey within the typically very confined space represented by each arena.

The general foraging arena assumption that predation typically is concentrated within restricted arenas, and hence at restricted rates, has profound implications for model predictions about ecosystem stability, and the further assumption that each predator-prey interaction takes place within a unique arena has equally profound implication for the maintenance of ecosystem structure and diversity[4]. It essentially represents the possibility of a distinct "feeding niche" for each of the predators that takes type i prey, hence allowing for the possibility that multiple predators can coexist while feeding on only that prey type. A prototype example of this possibility is with rockfishes (*Sebastes* spp.) along the Pacific coast, where a diverse collection of species all feed on euphausids, but avoid direct competition for these euphausids by feeding at different depths and times of day. An obvious evolutionary argument in favour of assuming such fine structure in feeding interactions is that if several predators were to feed within the same micro-scale foraging arena, the intense inter-specific competition caused by such behavior would result in very strong natural selection favouring differentiation of behaviors to avoid it, e.g., by feeding at different depths or times.

While there are evolutionary arguments in favour of assuming a distinct foraging arena for every interaction, Aydin and Gaichas[5] emphasize that there are some situations where multiple predator types are likely to feed on exactly the same prey and at the same place and time. An example could be where the predator types represent different life history stanzas (age-size classes) of the same predator species with very similar feeding modes (times and locations).

We represent this possibility in Ecosim by entry of base proportions of each predator type's diet that occurs in each of the possible foraging arenas defined by all non-zero predator-prey consumption linkages. Vulnerable prey density in each arena is then represented as varying over time in response to abundances of all predator types that feed in the arena.

In Ecosim, we define a list $a = 1, \ldots, N_a$ of possible foraging arenas, where N_a is the number of non-zero consumption interactions in the Ecopath diet matrix representing consumption of each prey type i by predator type j. Each of these potential arenas has a defining prey type $i(a)$ and defining predator type $j(a)$.

When only predator type $j(a)$ feeds in arena a, vulnerable prey density V_a is predicted by the basic foraging arena equation,

$$V_a = \frac{v_a \cdot B_{i(a)}}{v_a + v'_a + \alpha_a \cdot P_{j(a)}} \tag{1}$$

Here, v_a and v_a' are vulnerability exchange rates of prey to and from arena a, $B_{i(a)}$ is prey biomass, $P_{j(a)}$ is predator abundance (biomass or sum of numbers times search rates per predator for multi-stanza predators), and α_a is the predator rate of effective search (volume swept per time divided by foraging arena volume). The predation flow rate (biomass of prey $i(a)$ consumed per unit of time by predator $j(a)$) is then predicted as $Q_{i(a),j(a)} = \alpha_a V_a P_{j(a)}$. The v_a and α_a are parameterized by having model builders define va from maximum possible mortality rates expressed as multiples K_a of Ecopath base instantaneous predation rates $M_{ij}^{(0)} = Q_{ij}^{(0)} / B_i^{(0)}$, simply by setting $v_a = K_a$ where the superscript (0) designates Q's and B's estimated as base (initial) values of abundances and flows in the Ecopath baseline model. The back-exchange parameter v' is set equal to v since it cannot be estimated separately from the α_a parameter.

The shared-arena extension of Eq. 1 is straightforward,

$$V_a = \frac{v_a \cdot B_{i(a)}}{v_a + v'_a + \sum_k \alpha_{ak} \cdot P_k} \tag{2}$$

Here the predator impact on V_a is represented by a sum over all possible predators k of arena-specific search rates α_{ak} times predator abundances P_k. [In the software-implementation of this, we do not actually sum over all k but instead construct a list of all non-zero α_{ak} flow combinations, and sum the $\alpha_{ak}P_k$ denominator terms only over the elements of that list.]

To parameterize Eq. 2 in a relatively simple way while assuring that it predicts predation rates equal to Ecopath base rates when the system is at its Ecopath base state, we need to specify base proportions p_{ak} of each predator k's diet that is taken in arena a. These proportions are constrained to sum to Ecopath base consumption rates over all a for which $i(a) = i$. That is, we take the by-arena base flows to be $p_{ak} Q_{i(a),k}^{(0)}$. These base flows then imply a base instantaneous mortality rate $M_a^{(0)}$ totaled over predators feeding in a, for prey $i(a)$,

$$M_a^{(0)} = \frac{\sum_k Q_{ak}^{(0)}}{B_{i(a)}} \tag{3}$$

Using this input or baseline estimate of M for each arena and an assumed vulnerability multiplier K_a for that arena, we simply set $v_a = K_a$ (and $v_a' = v_a$).

Next, note that to be consistent with Ecopath baseline inputs, we must require that Ecosim predict $Q_{ak}^{(0)}$ when all biomasses (and p's) are at their Ecopath base values. The Ecosim prediction of rate Q_{ak} (flow rate of prey to predator k from feeding in arena a) at any time is $Q_{ak} = \alpha_{ak} V_a P_k$, implying we must constrain the α_{ak} so that $Q_{ak}^{(0)} = \alpha_{ak} V_a^{(0)} P_k^{(0)}$, i.e. we must set $\alpha_{ak} = Q_{ak}^{(0)} / (V_a^{(0)} P_k^{(0)})$. This means that to estimate the α_{ak} we must first estimate the base vulnerable abundances $V_a^{(0)}$.

This estimation turns out to be remarkably simple, when we note that the Ecopath base value of $\sum_k \alpha_{ak} P_k$ must equal $\sum_k Q_{ak}^{(0)} / V_a^{(0)}$, (simply sum Q_{ak} over k, which must equal $V_a = \sum_k \alpha_{ak} P_k$, and solve for $\sum_k \alpha_{ak} P_k$). Substituting $\sum_k Q_{ak}^{(0)} / V_a^{(0)}$ for $\sum_k \alpha_{ak} P_k$ in Eq. 2, then solving for $V_a^{(0)}$, we calculate the base vulnerable abundances to be simply,

$$V_a^{(0)} = v_a \cdot B_{i(a)} - \frac{\sum\limits_k Q_{ak}^{(0)}}{v_a + v_a'} \tag{4}$$

The α_{ak} are then calculated from these base vulnerable biomasses. Time-varying values of Q_{ak} are computed efficiently in Ecosim by setting up a list $h = 1, \ldots, N_h$ of all non-zero by-arena flows ($N_h \geq N_a$), where for each list element we store its associated prey type $i(h)$, predator type $k(h)$, and arena $a(h)$.

To calculate Q_{ak}, we sweep down this list repetitively. On the first sweep, we accumulate the denominator sums $\sum_k \alpha_{ak} P_k$ for Eq. 2. We then sweep down the arena list and calculate V_a for every a again using Eq. 2. Then we sweep again down the h list, calculating $Q_{ak} = \alpha_{ak} V_a P_k$ and accumulating predictions of total predation rates on the prey $i(a)$ and food consumption rates by predators $k(a)$.

As an added bit of model realism, one can specify a non-zero prey handling times for predator k (type II functional response[6]), and the Q_{ak} calculation is modified to be $Q_{ak} = (\alpha_{ak}/H_k) V_a P_k$, where H_k is the denominator of Holling's multi-species disc equation for predator k feeding. This handling time correction is also applied in the bout-feeding formulation described in the next chapter.

To edit the p_{ak} diet proportions array, we display a matrix for each prey type i of the non-zero i-k consumption proportions, as shown schematically in Table 1. In this table, m is the number of non-zero flows from prey i to predators k where each such flow defines a potential foraging arena. Note that each column of the table must sum to 1.0, i.e. all of the consumption by predator kj of prey type i must be accounted for by feeding in one of the m identifiable arenas for prey type i. The Ecosim default proportions for this table imply that each predator takes all of its consumption of prey type i in a unique arena, i.e. the table is an identity matrix, (with values of 1 on the shaded diagonal in Table 1).

Table 1 – Matrix showing linkages in the arena structure for a given prey with multiple predators (k_1 to k_m)

Arena	Predator			
	Predator k_1	Predator k_2	...	Predator k_m
Arena a_1	$p_{a1,k1}$	$p_{a1,k2}$	...	$p_{a1,km}$
Arena a_2	$p_{a2,k1}$	$p_{a2,k2}$		$p_{a2,km}$
...	...	...	...	...
Arena a_m	$p_{am,k1}$	$p_{am,k2}$	...	$p_{am,km}$

Each entry indicates the proportion (p_{ak}) of the activity for the given predator–prey interaction occurring in the corresponding arena a_i. In the default arena structure each predator is assumed to occupy a separate foraging arena indicated by the shaded cells on the diagonal. In such a case the diagonal will have a value of 1, and the other cells a value of 0. Each column must sum to 1.

The opposite extreme of this default assumption would be that all consumption of prey type i by its predators occur in only one arena or behavioral state for prey i, as shown in Table 2. This case implies maximum possible impact of predators k on availability of prey i to one another, and will cause competitive exclusion of at least some predator types in Ecosim unless the predators are well-differentiated in terms of overall diet composition, i.e. where each predator "specializes" on a different prey type i, which dominates the diet composition, as for instance shown by Schmidt[7]. Studies rather tend to indicate resource partitioning between competing predator species, leading to non-additive mortality rates, see e.g., Griffen and Byers[8]. Separation where diet compositions indicate predator overlap may also be caused by temporal exclusion of prey based on availability to the predator[9].

Table 2 – Example of an extreme arena structure where all predators (k_i) share a common foraging arena (a_1), i.e., compete fully for the same given prey

Arena	Predator			
	Predator k_1	Predator k_2	...	Predator k_m
Arena a_1	1.0	1.0	...	1.0
Arena a_2	0.0	0.0		0.0
...	...	...	...	...
Arena a_m	0.0	0.0	...	0.0

This will likely be an unstable situation with strong predator competition effects.

In the special case where a set of predators feeds on only one prey type in a single arena (Table 2), and where there are no complications such as multistanza population dynamics where abundance of one or more predator types may be limited by recruitment rates from younger stanzas, the above formulation implies that there is not even a unique equilibrium point for predator abundances. Rather, all predator abundance combinations that predict $V=V(0)$ in Eq. 2 are neutral stable points provided predator mortality rates remain at Ecopath base values, such that any temporary pulse of differential mortality that causes one or more predators to decline will then be followed by persistence of the new predator abundance combination if mortality rates return to the base values. Any predator that suffers a persistent differential increase in mortality rate is predicted to decline toward extinction.

Attribution This chapter was inspired by Kerim Aydin's work on a foraging arenas, and is based on Walters and Christensen. 2007.[10], used with permission from Elsevier, Licence Number 5663310244809.

Notes

1. Walters, C., V. Christensen and D. Pauly. 1997. Structuring dynamic models of exploited ecosystems from trophic mass-balance assessments. Reviews in Fish Biology and Fisheries 7:139-172. https://doi.org/10.1023/A:1018479526149

2. Walters, C.J., J.F. Kitchell, V. Christensen and D. Pauly. 2000. Representing density dependent consequences of life history strategies in aquatic ecosystems: Ecosim II. Ecosystems 3: 70-83. https://doi.org/10.1007/s100210000011

3. Christensen, V. and C. J. Walters. 2004. Ecopath with Ecosim: methods, capabilities and limitations. Ecol. Model. 172:109-139 https://doi.org/10.1016/j.ecolmodel.2003.09.003

4. Walters, C. J. and Martell, S. J. D., 2004. Fisheries Ecology and Management. Princeton University Press, Princeton. 399 pp.

5. Aydin, K. Y. and Gaichas, S. K., 2007. In defense of complexity: towards a representation of uncertainty in multispecies models. MS, SC/58/E, Alaska Fisheries Science Centre, NOAA, Seattle WA

6. Holling, C.S., 1959. The components of predation as revealed by a study of small mammal predation of the European pine sawfly 91, 293–320. https://doi.org/10.4039/Ent91293-5

7. Schmidt, K. A., 2004. Incidental predation, enemy-free space and the coexistence of incidental prey. Oikos, 106:335-343. https://doi.org/10.1111/j.0030-1299.2004.13093.x

8. Griffen, B. D. and Byers, J. E., 2006. Partitioning mechanisms of predator interference in different habitats. Oecologia, 146:608-614. https://doi.org/10.1007/s00442-005-0211-4

9. Scheuerell, J. M., Schindler, D. E., Scheuerell, M. D., Fresh, K. L., Sibley, T. H., Litt, A. H. and Shepherd, J. H., 2005. Temporal dynamics in foraging behavior of a pelagic predator. Canadian Journal of Fisheries and Aquatic Sciences, 62:2494-2501. https://doi.org/10.1139/f05-164

10. Walters, C and V. Christensen. 2007. Adding realism to foraging arena predictions of trophic flow rates in Ecosim ecosystem models: shared foraging arenas and bout feeding. Ecological Modelling 209:342-350. https://doi.org/10.1016/j.ecolmodel.2007.06.025

27.

Bout feeding

Many predators do not feed continuously over time as assumed in derivation of the vulnerable abundance Va in the standard foraging arena equation (Eq. 1 below). Rather, they obtain most of their food intake in short, intensive feeding "bouts", typically at dawn and dusk when light levels are changing rapidly[1][2].

Basic foraging arena equation

$$V_a = \frac{v_a \cdot B_{i(a)}}{v_a + v'_a + \alpha_a \cdot P_{j(a)}} \tag{1}$$

Shared foraging arena extension (see preceding chapter)

$$V_a = \frac{v_a \cdot B_{i(a)}}{v_a + v'_a + \sum_k \alpha_{ak} \cdot P_k} \tag{2}$$

Particularly when predators such as juvenile fish have severely restricted habitat use as a tactic for managing predation risk (hiding, schooling), only a small fraction of the system-scale prey biomass is available to them in the foraging arenas that they use during each feeding bout. As an example, juvenile Atlantic salmon have been shown to restrict the time they spend feeding rather than maximizing their growth when food is abundant[3].

Here, we show that overall trophic flow rates Q_{ak} (for predators k feeding in foraging arena a) over longer time scales can still be closely approximated by a continuous rate equation of the mass-action form $Q_{ak} = \alpha_{ak}V_{i(a)}P_{k(a)}$, where the bout search rates α^* and mean vulnerable prey densities per feeding bout are comparable to (but differ numerically from) the α,V predictions for continuous feeding.

Consider a single feeding bout in arena a of duration d ($d \ll$ one day), during which an initial prey density $V_a(0)$ is depleted by predators $k(a)$. Assume that d is short enough that prey renewal and loss during the bout, (e.g., due to prey spatial movement and other mortality sources) can be safely ignored. Assume that $V_a(0)$ is a proportion f_a of total prey biomass $B_{i(a)}$ and that renewal mechanisms between bouts make $V_a(0) = f_a B_{i(a)}$. Note that when used over multiple bouts, this prediction of $V_a(0)$ for each bout requires that arena prey abundance be independent of predation effects in previous bouts except through effects on $B_{i(a)}$, i.e. that there are no carryover effects from previous bouts (extreme opposite of continuous feeding assumption). Then if predators k search randomly within the arena, vulnerable prey density $V_a(t)$ will change during the bout according to the simple rate equation,

$$\frac{dV_a(t)}{dt} = -V_a(t) \sum_k \alpha_{ak} P_k \tag{3}$$

where the α_{ak} are predator rates of effective search with the same interpretation as for continuous feeding.

Integrating Eq. 3 over the bout duration d leads to the familiar exponential exploitation equation $V_a(d) = V_a(0) \exp(-d \sum_k \alpha_{ak} P_k)$ and to predicted total prey consumption per bout Q_{ak}^{bout} by each predator k,

$$Q_{ak}^{bout} = \frac{\alpha_{ak} P_k}{\sum_k \alpha_{ak} P_k} \cdot f_a B_{i(a)} \cdot [1 - \exp(-d \sum_k \alpha_{ak} P_k)] \tag{4}$$

The first term of Eq. 4 simply apportions total prey consumption $V_a(0) - V_a(d)$ over the bout among competing predators. Further, the mean prey density V_a^* during the bout is given by the integral of V over the bout divided by bout duration d. This mean is just,

$$V_a^* = f_a B_{i(a)} \cdot \frac{1 - \exp(-d \sum_k \alpha_{ak} P_k)}{d \sum_k \alpha_{ak} P_k} \tag{5}$$

Expressed in terms of this mean arena prey density, consumption per bout Eq. 4 can be expressed more simply as

$$Q_{ak}^{bout} = \alpha_{ak} P_k V_a^* \tag{6}$$

We could use this formula directly in a complex simulation model that steps forward in time by the interval Δ_t between feeding bouts, adding in other components of prey and predator abundance change over each such short interval. Fortunately, such a tedious calculation is generally unnecessary.

Consider the component of overall prey biomass change caused by each feeding bout, where there are $n_b = 1/\Delta_t$ bouts per year. That (typically small) change in $B_{i(a)}$ per bout is given by the sum of Eq. 4 terms over predators k, i.e.,

$$\Delta B_{i(a)} = f_a B_{i(a)} \cdot [1 - \exp(-d \sum_k \alpha_{ak} P_k)] \tag{7}$$

Dividing this by the bout duration Δ_t gives a discrete-time component of the prey rate of change,

$$\frac{\Delta B_{i(a)}}{\Delta t} = \frac{1}{\Delta t} f_a B_{i(a)} \cdot [1 - \exp(-d \sum_k \alpha_{ak} P_k)] = n_b f_a B_{i(a)} \cdot [1 - \exp(-d \sum_k \alpha_{ak} P_k)] \tag{8}$$

Since the time Δ_t between bouts is typically very short (n_b is typically of the order of several hundred bouts per year), we can approximate Eq. 8 very accurately by treating it as a continuous rate component $dB_{i(a)}/dt$. This approximation leads immediately to a continuous rate equation for Q_{ak} comparable to the continuous feeding case where $Q_{ak} = \alpha_{ak} P_k V_a$, namely,

$$Q_{ak} = \alpha_{ak}^* \, P_k \, v_a^* \, B_{i(a)} \cdot \frac{1 - \exp(-\sum_k \alpha_{ak}^* P_k)}{\sum_k \alpha_{ak}^* P_k} = \alpha_{ak}^* \, P_k \, V_a^* \tag{9}$$

Where α_{ak}^* are the duration-weighted search rates $\alpha_{ak}^* = \alpha_{ak}d$. $v_a^* = n_b \, f_a$ represents a total prey "fraction" that would become vulnerable over a one-year time scale, and V_a^* (comparable to Eq 2) is given by

$$V_a^* = v_a^* \, B_{i(a)} \cdot \frac{1 - \exp(-\sum_k \alpha_{ak}^* \, P_k)}{\sum_k \alpha_{ak}^* \, P_k} \tag{10}$$

This model for vulnerable prey density obviously exhibits the same "ratio dependence" of available prey density on predator abundance as does Eq 2, but with the ratio effect $1/(v+v'+\sum_k \alpha_{ak} P_k)$ replaced by a negative exponential effect. At high predator abundances it also implies an upper bound $B_{i(a)}$ on total removal rate Q_a and hence on total instantaneous predation mortality rate $Q_a/B_{i(a)}$.

We can parameterize the continuous approximation to bout feeding from Ecopath inputs and assumed maximum predation rates in the same way as described in the previous section for continuous arena feeding. That is, we set $v_a^* = K_a \, M_a(0)$ where K_a as above is a defined input ratio of maximum to Ecopath baseline predation rate. We calculate base mean prey density per bout $V_a^*(0)$ by substituting Ecopath base prey and predator abundances $B_{i(a)}(0)$ and $P_k(0)$ into Eq. 10 along with $\sum_k \alpha \, P_k(0) = Q_a(0) / V_a(0)$, (where $Q_a(0)$ is the base total consumption rate summed over predators k), and solving for V_a^*, to give

$$V_a^*(0) = -\frac{Q_a(0)}{\ln(1 - \frac{1}{K_a})} \tag{11}$$

Then we simply calculate the α_{ak}^* as

$$\alpha_{ak}^* = -\frac{Q_{ak}(0)}{P_k(0) \cdot V_a^*(0)} \tag{12}$$

whereas above the arena-specific base consumption rate is calculated using assumed arena feeding proportions p_{ak} as $Q_{ak}(0) = p_{ak}Q_{i(a),k}(0)$, and $Q_{i(a),k}(0)$ is the Ecopath base total consumption rate of prey $i(a)$ by predator k.

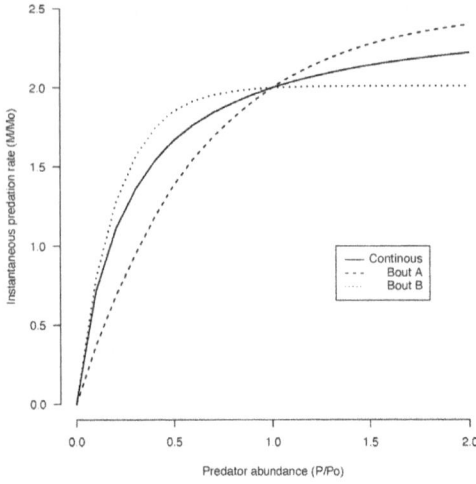

Figure 1. Comparison of instantaneous mortality rates expressed relative to Ecopath baseline predation mortality rates for the original model formulation ("Continuous") and compared to bout feeding with $v_a^* = v_a$ in case *Bout A*, and with $v_a^* < v_a$ set to give same limiting maximum consumption per predator, Q/P in case *Bout B*.

It is instructive to compare the predictions of instantaneous prey mortality rate $M = Q_a/B_{i(a)}$ from Eq. 8 to those of the continuous model defined by Eq. 2 (and $Q_a = \sum_k \alpha_{ak} P_k V_a$), for varying predator abundances P_k while holding prey biomasses $B_{i(a)}$ constant (Figure 1). If we set $v_a^* = v_a$, i.e. use the same K_a to calculate v_a^* as we would for v_a in the continuous case (*Bout A* in Figure 1), the exponential term in the bout feeding model generally predicts steeper variation in M than the continuous model, i.e., it predicts that M will drop off more rapidly if P decreases from $P(0)$ than does the continuous model. This leads to weaker "compensation" measured in terms of increase in potential Q/P as P declines. But if we set v_a^* smaller than v_a, so as to predict the same limiting maximum consumption per predator (Q/P) at very low predator densities (*Bout B* in Figure 1), the two arena models give predicted patterns of variation in M that are the opposite, i.e. bout feeding predicts saturation of M at lower P than the continuous case. This means that in Ecosim cases where K_a has been estimated by fitting the continuous arena model to time series data (the only option before the inclusion of bout feeding in EwE), and where feeding in reality has been of the bout type, the fitted K_a estimates have probably been somewhat too large, i.e K_a is in reality closer to 1.0 and predators are already causing (in the Ecopath base situation) what may be close to their maximum possible predation rates from bout feeding.

For Ecosim models that include multi-stanza population dynamics, a critically important capability is to represent adjustments in foraging time, particularly for juvenile stanzas. Such adjustments allow juvenile fish to translate increases in potential feeding rate Q/P into reduced foraging time and predation risk when competitor abundance P decreases[4], leading to compensatory changes in juvenile mortality rates and emergent stock-recruitment relationships of the Beverton-Holt form[5].

Foraging time adjustments are modeled in Ecosim by including a dynamic variable T_i for each biomass type, with T_i at time zero set to 1.0. Then T_i is varied over time so as to try and maintain Ecopath base feeding rate per predator (Q/P), by multiplying all search rates of type i for its prey by T_i, and all vulnerability exchange rates of type i into arenas where predators take it by T_i. In the bout foraging representation, this means simply that

(1) search parameters for type i as a predator are adjusted by varying bout durations d in proportion to T_i (i.e. setting $\alpha^*(t) = \alpha^*(0)\, T_i$, with T_i defined as the relative bout duration $d(t)/d(0)$) and (2) the vulnerable fraction f that define v^* of i to its predators are also treated as being proportional to d by setting $v^*(t) = v^*(0)\, T_i$.

As a simple test of whether bout feeding is likely to make much difference to the ability of Ecosim models to fit historical time series data, we examined changes in a simple fitting criterion (sum of squared deviations, SS, from historical data, summed over all time series used in model fitting) for a collection of models that had been fitted to data using the continuous arena equations, when all trophic linkages were simply reset to assume bout feeding without correction or refitting of the K_a parameters (Figure 2,[6]). Ability of several of these models to fit historical data are reviewed in Walters and Martell (2004, Figure 12.6[7]), and these were mostly the same models used in single-species versus multispecies MSY comparisons by Walters et al.[8].

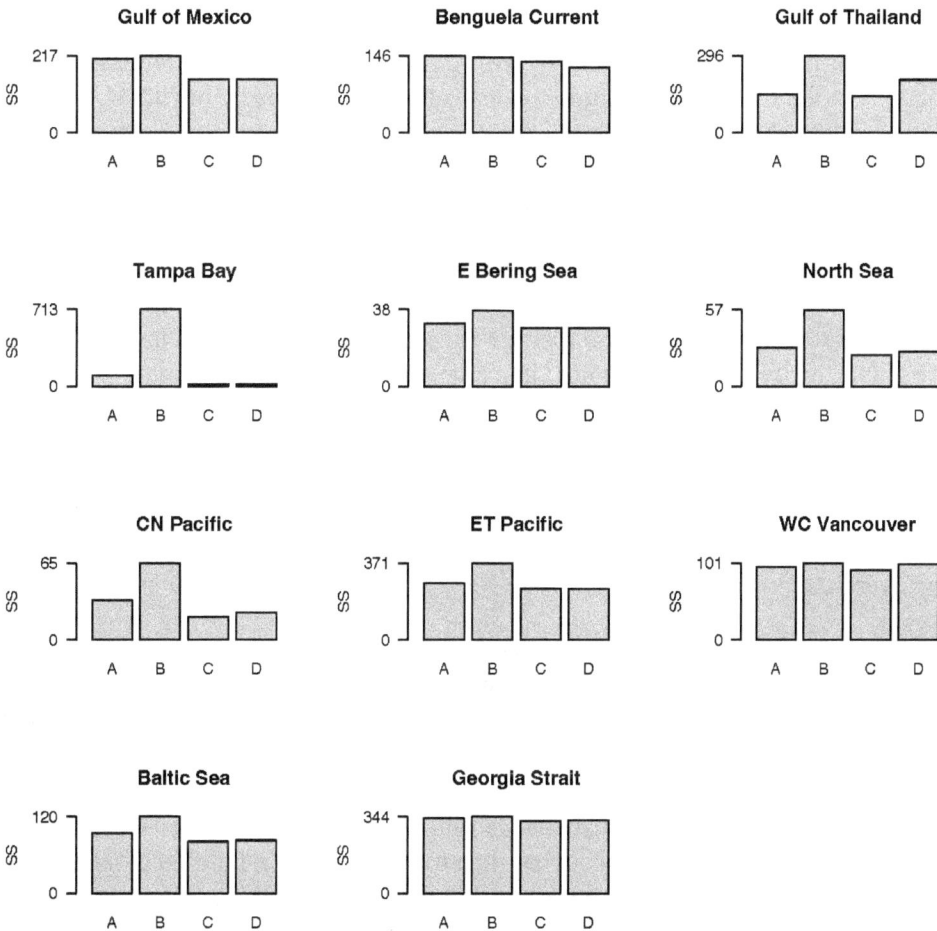

Figure 2. Changes in sum of squares goodness-of-fit criterion (SS) for an assort-

ment of models that have been fit to historical data using the continuous feeding arena equations of Ecosim, when bout feeding is assumed instead for all trophic linkages initially without re-estimation of vulnerability exchange multipliers K_a. **(A): represents *SS* from original model. (B): *SS* from original model but with bout feeding. (C): *SS* for bout feeding, after fitting by varying 20 most important *Ka*. (D): *SS* for continuous feeding after fitting.**

Surprisingly, there was little change in the fitting criterion for many of the models, and one (Central North Pacific) even gave a better fit immediately. For those models where there was a substantial increase in SS, it was easy to remedy the poor fits by re-estimating K_a under the global bout arena assumption. When we refitted the models under both feeding assumptions (by nonlinear estimation search over the 20 K_a values with largest contributions to the sum of squares), we were easily able to find fits at least as good under the bout feeding assumption for most cases, and qualitatively as good for all cases.

Discussion

The equations introduced here obviously give considerable flexibility to represent trophic interactions in the Ecosim model more realistically than previously possible[9]. It is particularly comforting to see that the much more realistic assumption of bout rather than continuous feeding leads to very similar predictions of how prey mortality rates should vary with predator abundances as have been assumed in previous Ecosim models based on the unrealistic but mathematically convenient assumption of continuous feeding with rapid equilibration of vulnerable prey densities.

We recommend extreme care in using either the continuous or bout feeding equations to represent feeding by multiple predators in a relatively small number of arenas. As noted above, the intense inter-specific competition implied by such concentration of feeding has very likely driven natural selection for differentiation in feeding behavior (use of different fine-scale arenas) as well as in diet composition. See for instance Berec *et al.* (2006) for an experiment illustrating this. If such differentiation is excluded from the model parameterization, the Ecosim user risks building a model that will not retain observed biodiversity over time.

A few authors have referred to the basic Ecosim equation for predicting total flow rates Q $=(\alpha\, v\, B\, P) / (v + v' + \alpha\, P)$ as though it were a functional response equation comparable to assuming mass-action encounters and type II predation, e.g., $Q =(\alpha\, B\, P) / (1 + h\, B)$;[10]. Such comparisons reflect a misunderstanding about a basic proposition of foraging arena theory, namely that predators very generally encounter their prey in space-time restricted circumstances (foraging arenas), such that it is almost never appropriate to predict Q from the ecosystem-scale mean prey density B when trying to account for effects such as handling time and switching (changes in α). We would argue that it is sometimes appropriate to account for handling time effects, but only if these are predicted using arena-scale vulnerable prey densities V, i.e. $Q = \alpha\, V\, P / (1 + h V)$, where V is adjusted away from the system-scale average B using assumptions about localization of foraging (effects of vulnerability exchanges v's and/or available prey fractions per bout f's in the arena equations). We explicitly allow switching in Ecosim, but again caution that it should be used in conjunction with predictions of vulnerable, rather than overall, prey densities.

It would be ignorant to assert that the equations presented in this chapter are the only or best way to represent differentiation of vulnerable prey biomasses V from system-scale average prey biomasses B in prediction of trophic interaction rates. They do not for example account explicitly for some very gross system-scale effects that occur in highly disturbed systems, such as changes in overall prey and predator distributions and overlap patterns, (e.g., due to range contractions), and changes in spatial arena structure due to obvious habitat changes like growth and destruction of biogenic spatial refuges[11]. Some such changes can be accounted for in Ecosim through trophic 'mediation functions' that link v's and α's to abundances of species besides those engaged directly as predators and prey, (e.g., one can make α's and v's for juvenile fish that hide in macrophyte beds dependent on macrophyte biomass). But there is still a long way to go in development of fully-defensible predictions of V for systems that are massively disturbed.

One option for dealing with the prediction of V would be to construct very detailed spatial models (with habitat and its use modelled at fine scale, maybe of a few m^2) running on very short (bout) time scales (time steps of one hour or less). But such models may be plagued by lack of detailed spatial data, lack of understanding of how organisms move and concentrate their activities at such fine scales, and risk of cumulative divergence of predictions from reality simply due to explosions over simulated time and space of small errors in behavioural movement predictions.

A key advantage of the relatively simple foraging arena equations for Q prediction is that we can easily force them to agree with baseline "observations" or estimates of system-scale abundances (B's, P's), feeding rates, and diet compositions (Q_{ji}) as summarized in static (point-in-time) mass-balance assessments like Ecopath. But this is also a disadvantage, in the sense that the rate parameter estimates then become dependent on the often incomplete and possibly biased estimates entered as Ecopath inputs. It is clear that Ecosim-type dynamic predictions are sensitive to those baseline inputs, and that this represents an especially severe issue for interactions involving small fish as prey (where the small fish typically represent only trivial and often overlooked proportions of their predators' diets).

Even absent difficulties with Ecopath inputs, i.e. empirical knowledge of baseline ecosystem biomass flow rates and states, the most troublesome parameters for Ecosim users to specify have been the "vulnerability multipliers" K_a representing ratios of maximum to Ecopath base predation mortality rates. One source of trouble is obviously that Ecopath inputs provide no information about the K_a, and such information can only come from either fine-scale analysis of spatial arena structures, from data collected at different times and/or places about how Q's have varied with predator and prey abundances, or from assumptions about or estimates of where populations are relative to their carrying capacity. Indeed, this is why we emphasize the importance of fitting Ecosim models to time series data by varying the Ka parameters.

Another, and important aspect is that the K_a are not purely "behavioural" or ecological parameters; rather, they depend as well on how large the Ecopath initial predator abundances P_k are compared to what the ecosystem might naturally support (see Density dependence chapter). So for example a model that includes Atlantic cod stocks off Newfoundland, and uses the current low stock size as the Ecopath base, must have very high K_a values (1000+) for interactions between cod and its prey, else the model will not make enough prey available to the simulated cod stock for it to recover to anywhere near its historical abundance when simulated fishing is removed.

We can provide some guidance about reasonable ecological Ka values (corrected for effects of historical depletion on biomasses) from meta-analysis of K_a estimates for many fitted models. One pattern that is becoming broadly evident from cases like those in Figure 2 is that fitted K_a values tend to be small (<2.0) for most trophic linkages in temperate and tropical systems, and for feeding by juvenile stanzas in all systems. In contrast, fitted K_a values tend to be much larger for most interactions (except juvenile stanzas of demersal fish species) in high-latitude ecosystems like the Bering Sea. The low K_a values are easily explained for juvenile fish and reef-associated older fish, as a consequence of severe spatial restriction in habitat use leading to low proportions of prey populations being available the fish at any time[12]. High K_a values in high-latitude systems likely reflect the wider spatial movement characteristic of northern fish, and tactics such as diel vertical migration that bring high proportions of widely distributed predators and prey into daily contact with one another, (i.e., high f's for bout feeding during periods of diurnal contact[13]).

The K_a vulnerability multipliers are discussed in details in the vulnerability multiplier chapter.

Attribution This chapter is based on Walters and Christensen (2007)[14], used with permission from Elsevier, Licence Numbers 5663310244809 and 5663310474242.

Notes

1. Helfman, G. S., 1993. Fish behaviour by day, night and twilight. In: T. J. Pitcher (Editor) Behaviour of Teleost Fishes. Chapman & Hall, London, Vol. 2. pp. 479-512.

2. Rickel, S. and Genin, A., 2005. Twilight transitions in coral reef fish: the input of light-induced changes in foraging behaviour. Animal Behaviour, 70:133-144. https://doi.org/10.1016/j.anbehav.2004.10.014

3. Orpwood, J. E., Griffiths, S. W. and Armstrong, J. D., 2006. Effects of food availability on temporal activity patterns and growth of Atlantic salmon. Journal of Animal Ecology, 75:677-685. https://doi.org/10.1111/j.1365-2656.2006.01088.x

4. see, e.g., Orpwood, J. E., Griffiths, S. W. and Armstrong, J. D., 2006. Effects of food availabil-

ity on temporal activity patterns and growth of Atlantic salmon. Journal of Animal Ecology, 75:677-685. https://doi.org/10.1111/j.1365-2656.2006.01088.x

5. Walters, C. and Korman, J., 1999. Linking recruitment to trophic factors: revisiting the Beverton-Holt recruitment model from a life history and multispecies perspective. Reviews in Fish Biology and Fisheries, 9:187-202. https://doi.org/10.1023/A:1008991021305

6. for details about the models, see Walters, C and V. Christensen. 2007. Adding realism to foraging arena predictions of trophic flow rates in Ecosim ecosystem models: shared foraging arenas and bout feeding. Ecological Modelling 209:342-350. https://doi.org/10.1016/j.ecolmodel.2007.06.025

7. Walters, C. J. and Martell, S. J. D., 2004. Fisheries ecology and management. Princeton University Press, Princeton. 399 pp.

8. Walters, C. J., Christensen, V., Martell, S. J. and Kitchell, J. F., 2005. Possible ecosystem impacts of applying MSY policies from single-species assessment. ICES Journal of Marine Science, 62:558-568. https://doi.org/10.1016/j.icesjms.2004.12.005

9. Walters, C., Pauly, D., Christensen, V. and Kitchell, J. F., 2000. Representing density dependent consequences of life history strategies in aquatic ecosystems: EcoSim II. Ecosystems, 3:70-83. https://doi.org/10.1007/s100210000011

10. see, e.g., Koen-Alonso, M. and Yodzis, P., 2005. Multispecies modelling of some components of the marine community of northern and central Patagonia, Argentina. Canadian Journal of Fisheries and Aquatic Sciences, 62:1490-1512. https://doi.org/10.1139/f05-087

11. e.g., Rodriguez, C. F., Becares, E., Fernandez-Alaez, M. and Fernandez-Alaez, C., 2005. Loss of diversity and degradation of wetlands as a result of introducing exotic crayfish. Biological Invasions, 7:75-85. https://doi.org/10.1007/s10530-004-9636-7

12. e.g., Gonzalez, M. and Tessier, A., 1997. Habitat segregation and interactive effects of multiple predators on a prey assemblage. Freshwater Biology, 38:179-191.

13. see, e.g., Hrabik, T. R., Jensen, O. P., Martell, S. J. D., Walters, C. J. and Kitchell, J. F., 2006. Diel vertical migration in the Lake Superior pelagic community. I. Changes in vertical migration of coregonids in response to varying predation risk. Canadian Journal of Fisheries and Aquatic Sciences, 63:2286-2295. https://doi.org/10.1139/f06-12

14. Walters, C and V. Christensen. 2007. Adding realism to foraging arena predictions of trophic flow rates in Ecosim ecosystem models: shared foraging arenas and bout feeding. Ecological Modelling 209:342-350. https://doi.org/10.1016/j.ecolmodel.2007.06.025

V

Fitting Models to Data

28.

Density-dependence, carrying capacity and vulnerability multipliers

An important feature of Ecosim is that it provides a straightforward approach for exploring alternative views for how the biomass of prey groups are controlled by a predator – or if they are. The two extreme views are "predator" control (also called "top-down") and "prey control" (or "bottom-up") – concepts that are quite difficult to fully grasp. But they become really important when you start comparing a model to historical data as a way of establishing credibility of the model's predictions, because assumptions about them dramatically impact whether the model can "track" or reasonably represent known historical disturbances such as increase in fishing mortality rates. Thus our main attention in both qualitative and quantitative methods for comparing EwE models to data is at least initially on the parameters that determine the pattern of trophic control.

We model this interplay between predator and prey control using "vulnerability multipliers," which provide a factor for how much an increase in predator abundance may impact the predation mortality that it is causing on a given prey.

- Low vulnerability multipliers (close to 1) mean that an increase in predator biomass will not cause any noticeable increase in the predation mortality the predator may cause on the given prey (see Figure 3.1).

- High vulnerability multipliers, (e.g., 100), indicates that if the predator biomass is for instance doubled, it will cause close to a doubling in the predation mortality it causes for a given prey.

Figure 1. Lesser flamingos in Lake Nakuru, Kenya. Millions of flamingos overwinter in the lake where they feed on brine shrimp (Artemia) in the shallow parts of the lake. Do flamingos control the prey population? or is it the productivity of the Artemia population that sets how much the flamingos get to eat? Would twice as many flamingos eat twice as many Artemia?

On carrying capacity

When a predator is at its carrying capacity, its production depends on how productive its prey populations are. If it's a good year in the environment with high prey production, there will be more food for the predator – and its carrying capacity may increase. *Vice versa* if productivity is low. How much a predator population gets to eat depends on how productive their prey populations are. So, when a predator is close to its carrying capacity the system is "bottom-up" controlled. In this case, the Ecosim vulnerability multipliers are low, close to 1.

If the predator is far from its carrying capacity, the predator has very little control over its prey populations (there are too few of them!), but it's called "top-down" control because how much the predators eat depends on how many predators there are (rather confusing, eh?) Twice as many predators may eat twice as much food (and hence cause twice as high predation mortality) but have no notable impact on the prey population (still, it's called "top-down" control). Here, we need to use high vulnerability multipliers in Ecosim.

You can think of it like this. The vulnerability multiplier sets how many times the predation mortality a predator is causing on its prey may increase if the predator population were to increase to its carrying capacity. If that multiplier is 100, then the predator is far from its carrying capacity – the population can perhaps grow >100 times before reaching carrying capacity. ">100" because the competition between the predators will be intense when they are at carrying capacity, so the average individual will get less food than if their population was much lower.

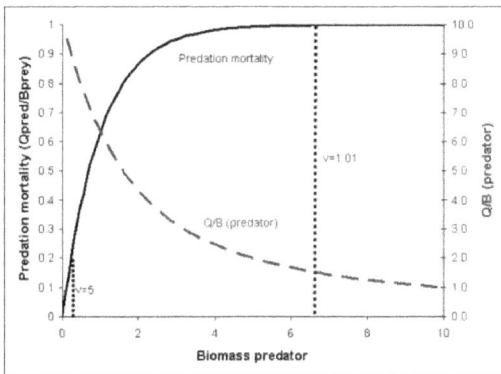

Figure 2. Relationship between biomass of a predator and the predation mortality it causes on a given prey, as well as the corresponding Q/B for the given predator and prey (assuming that the predator does not reduce prey biomass substantially). Vulnerability multipliers, v, are estimated as max. predation mortality/baseline predation mortality, (e.g., 5 at the left-most stippled line). Baseline mortality is the mortality caused by the predator in the underlying Ecopath model.

If we illustrate the relationship between predator biomass and Q/B (this is not an assumption in the actual Ecosim calculations) and assume that the predator in question does not cause any substantial (actually no) change in the prey biomass, we can calculate the relative Q/B for the predator (see Figure 2). For higher predator biomass, a change will result in relatively stable predation mortality. Hence, if biomass is impacted so as to cause a reduction, the individual predators will get more, their Q/B increase and this will largely compensate for the reduction in their abundance, bringing the biomass back up again.

At lower biomass, Q/B will also increase, but to a lower degree. This is illustrated in Figure 3 showing how halving or doubling the predator biomass will impact the relative Q/B. At high biomasses, halving biomass results in close to a doubling in Q/B, which will tend to keep biomass high. There is, however, less and less relative surplus production as we move to the left on the curve. If biomasses are doubled instead, the Q/B will decrease when biomasses are high, resulting in a decrease in biomass back toward the original level, i.e., the biomasses will be stable when close to carrying capacity (where v's are low), and unstable when far below carrying capacity (where v's are high).

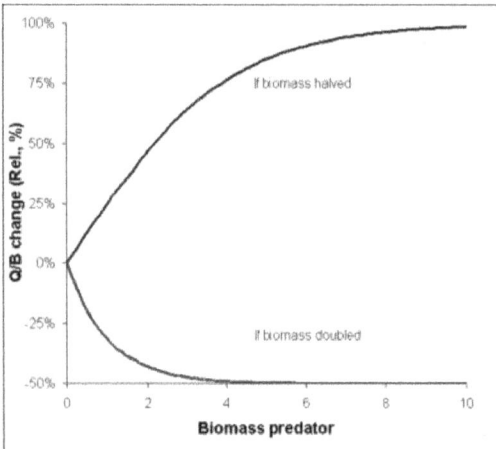

Figure 3. Relative increase (%) in Q/B as a function of predator biomass resulting from the predator biomass being halved or doubled. At high predator biomasses (i.e. near the carrying capacity for the given predator-prey interaction) a halving of predator biomass will result in nearly a doubling in the Q/B for the predator. The resulting surplus production will tend to bring the predator biomass back to the original level, and the overall effect is that the predator biomass will change only little. Conversely, a doubling of predators will cause the Q/B to be halved at high predator biomasses (resulting in very little effective change in biomass), while a doubling at low biomasses will result in only a very small reduction in Q/B.

If vulnerabilities are high, the amount of prey consumed by the predator is the product of predator times prey biomass, i.e., the predator biomass directly impacts how much of the prey is consumed. Such situation may occur in a situation where the prey has no refuge, and is thus always taken upon being encountered by a predator. Such top-down control, also known as Lotka-Volterra dynamics, easily leads to rapid oscillations of prey and predator populations and with it, making it impossible to maintain prey populations in a model.

Lotka-Volterra dynamics imply that a predator's consumption equals number of predators times number of prey times a search rate factor.

More predators means more consumption, and more prey means more consumption.

In Ecosim, however, top-down control implies that a predator is far from its carrying capacity. Hence, if there's only a small predator population around and conditions (available food, mortality reduction) allow it, the predator population may grow towards its carrying capacity. How much it may grow, given favourable conditions, depends on the vulnerability multiplier.

If you are modelling an area where a predator population is close to its carrying capacity – maybe something like Kingman Reef where apex predators make up 85% of the total fish biomass – then those predators depend on how productive the prey population is. More predators does not mean more prey consumption, the system is bottom-up controlled, and vulnerability multipliers should be low for such predators. This is a stable ecosystem configuration. If a burst of fishing should occur at

Kingman Reef, those fish that are left will see improved prey conditions, their Q/B will increase, and the population will grow back towards its carrying capacity.

Bottom-up control can also occur where a prey is protected most of the time, (e.g., by hiding in crevices) and becomes available to predators only when it leaves the feature that protects it. Here being caught is a function of the prey's behaviour. Bottom-up control implies stable system conditions, so it's associated with only small biomass changes in the prey and predator(s) concerned for instance as a function of fishing pressure.

The converse (top-down control) is the situation that occurs when a predator population is far from its carrying capacity – for instance because the population has been fished to the brink. In this situation, there are only few predators around and their prey populations may have increased due to predator release or cascading.

To model this interplay between top-down and bottom-up control in predator-prey interactions, the group biomasses in the underlying Ecopath model were in the foraging arena theory[1][2][3][4] in Ecosim conceived as consisting of two components, one vulnerable and one invulnerable to predation. We'll discuss this in more detail in the next chapter.

Model behaviour depends strongly on the vulnerability multipliers. How do you set those then? The most common way is to fit the model to time series data, which may imply fitting to vulnerabilities along with environmental productivity (see chapter). It worth remembering though that the vulnerability multipliers are not "nuisance parameters" (i.e. parameters which have no specific interpretation), but express how far a predator population is from its carrying capacity

Vulnerability multiplier = how many times might this predator increase the predation mortality it's causing on its prey, if it were to grow to its carrying capacity?

Carrying capacity is not constant; in Ecosim the vulnerability multipliers are relevant for and used only in the Ecopath baseline. Carrying capacity may then change for each and every time step in Ecosim if need be. Ecosim considers that and populations may change accordingly.

If you reflect on figures such as Figures 2 and 3, note that the x-axis is biomass. That implies we are talking about density-dependence, how much impact a predator has on its prey populations and how much a predator gets to eat is density-dependent.

Quiz

An interactive H5P element has been excluded from this version of the text. You can view it online here:

https://pressbooks.bccampus.ca/ewemodel/?p=996#h5p-2

Media Attributions

- Large_number_of_flamingos_at_Lake_Nakuru adapted by Licensed under the Creative Commons Attribution-Share Alike 3.0 Unported, 2.5 Generic, 2.0 Generic and 1.0 Generic license.

Notes

1. Walters, C. J. and F. Juanes (1993). Recruitment limitations as a consequence of natural selection for use of restricted feeding habitats and predation risk taking by juvenile fishes. Can. J. Fish.Aquat. Sci. 50, 2058-2070

2. Walters, C. J. and J. Korman (1999). Linking recruitment to trophic factors: revisiting the Beverton-Holt recruitment model from a life history and multispecies perspective. Rev. Fish Biol. Fish. 9, 187-202.

3. Walters CJ, SJD Martell. 2004. Fisheries Ecology and Management. Princeton University Press.

4. Ahrens, R. N. M., Walters, C. J., and Christensen, V. 2012. Foraging arena theory. Fish and Fisheries 13: 41–59.

29.

Vulnerability and vulnerability multipliers

Jacob Bentley; David Chagaris; Marta Coll; Sheila JJ Heymans; Natalia Serpetti; Carl J. Walters; and Villy Christensen

Ecosim predictions are sensitive to the Ecopath input parameters (usually biomass, production and consumption rates, diet, and fishery removals) as well as the predator-prey "vulnerability multipliers", which are conditioned on the Ecopath inputs and tuned during model calibration. In Ecosim, vulnerability multipliers have implications for stock-recruit dynamics, density dependence and compensation, carrying capacity, stock resiliency, interspecific interactions, and ecosystem energy flow[1]. However, the effect of different calibration strategies on the estimation of vulnerability multipliers in Ecosim and emergent stock productivity estimates has not yet been demonstrated, nor do we understand in a comprehensive way how sensitive model outputs are to different approaches and how this may influence derived advice.

Ecosim predictions of consumption based on a simple mass-action model have been modified to consider the non-random dynamics of the foraging arena[2][3][4]. Prey biomass pools in Ecosim are dynamically divided into vulnerable and invulnerable components, which imply behavioural or physical mechanisms that limit the rate at which prey become vulnerable to predation (Figure 1[5]). The transfer rates between these components determine the amount of prey available to a predator and thus the degree to which a change in predator biomass will impact predation mortality and prey biomass.

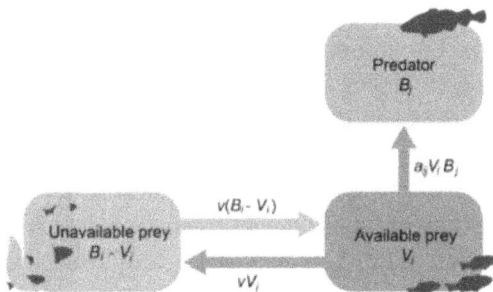

Figure 1. Simulation of flow between available (V_i) and unavailable ($B_i−V_i$) prey biomass in Ecosim. a_{ij} is the search rate of prey i by predator j, v is the exchange rate between the vulnerable and un-vulnerable state. Fast equilibrium between the two prey states implies $V_i=vB_i/(2v+aBj)$. Based on Walters et al. (1997).

These transfer rates, or "vulnerabilities" (v_{ij}) as they are more widely known, influence predator and prey biomasses by regulating the consumption rates (Q_{ij}) of a predator j as described in the Ecosim predicting consumption chapter.

When working with Ecosim, one cannot adjust vulnerability exchange rates (v_{ij}) directly.

Instead, this is done via vulnerability multipliers (k_{ij}), which can be more easily interpreted as the maximum increase in predation mortality rate that a predator can exert on a prey if the predator were to grow to its carrying capacity. The increase is relative to baseline Ecopath predation mortality rates ($M2$, where $M2=Q_{ij}/B_i$) The vulnerability exchange rates (v_{ij}) are then set to the vulnerability multiplier (k_{ij}) multiplied by the baseline predation mortality ($M2$), i.e. $v_{ij}=k_{ij}M2$. Multipliers can range from one to infinity with two as the default value.

The vulnerability multipliers are derived from the Ecopath baseline, and do not automatically change when running Ecosim across years. Ecosim dynamically handles the consequences of changes in predator and prey abundance based on the baseline situation, including changes in carrying capacity over time. The default value for k_{ij} of 2.0 assumes that the predation mortality rate can double at most, while a value down near 1.0 means that the predator is at its "carrying capacity", which by definition means it fully utilizes its prey, so it cannot further increase the predation mortality it's causing on the prey.

High vulnerability multipliers imply top-down control, and low bottom-up control. Top-down control occurs where a predator is far from its carrying capacity, here, e.g., a doubling of predator abundance may result in close to a doubling in the predation mortality it are causing on it prey. With low vulnerability multipliers where a predator is at its carrying capacity, any increase in consumption has to be linked to changes in prey productivity – i.e. to bottom-up factors.

For exploited species, it is thus extremely important to recognize that vulnerability multipliers do not only reflect the ecological limits caused by prey and predator behaviour, but also how depleted the exploited species is in the base Ecopath biomass state relative to the natural level (i.e., carrying capacity) that might be achieved if fishing were stopped. As such, for overexploited species to recover following reduced fishing, vulnerability multipliers need to be set relatively high so that predators can consume far more prey than in the initial Ecopath snapshot. Higher vulnerability multipliers tend to make groups more sensitive and responsive to changes in fishing mortality.

Over time, multiple approaches to parameterize the vulnerability multipliers have been developed and adopted (Figure 2). While some approaches derive estimates from *a priori* knowledge and ecological observations or hypotheses in data-poor situations[6], users are more frequently turning to formal statistical estimation using calibration time series and a tuning process when time series are available[7]. Statistical fitting routines estimate vulnerability multipliers that bring simulations closer in-line with observations. However, users should be cautious, as statistically optimized multipliers may stray away from values that might be considered ecologically realistic. A thorough sense check is always recommended. The following sections explore these different approaches in more detail.

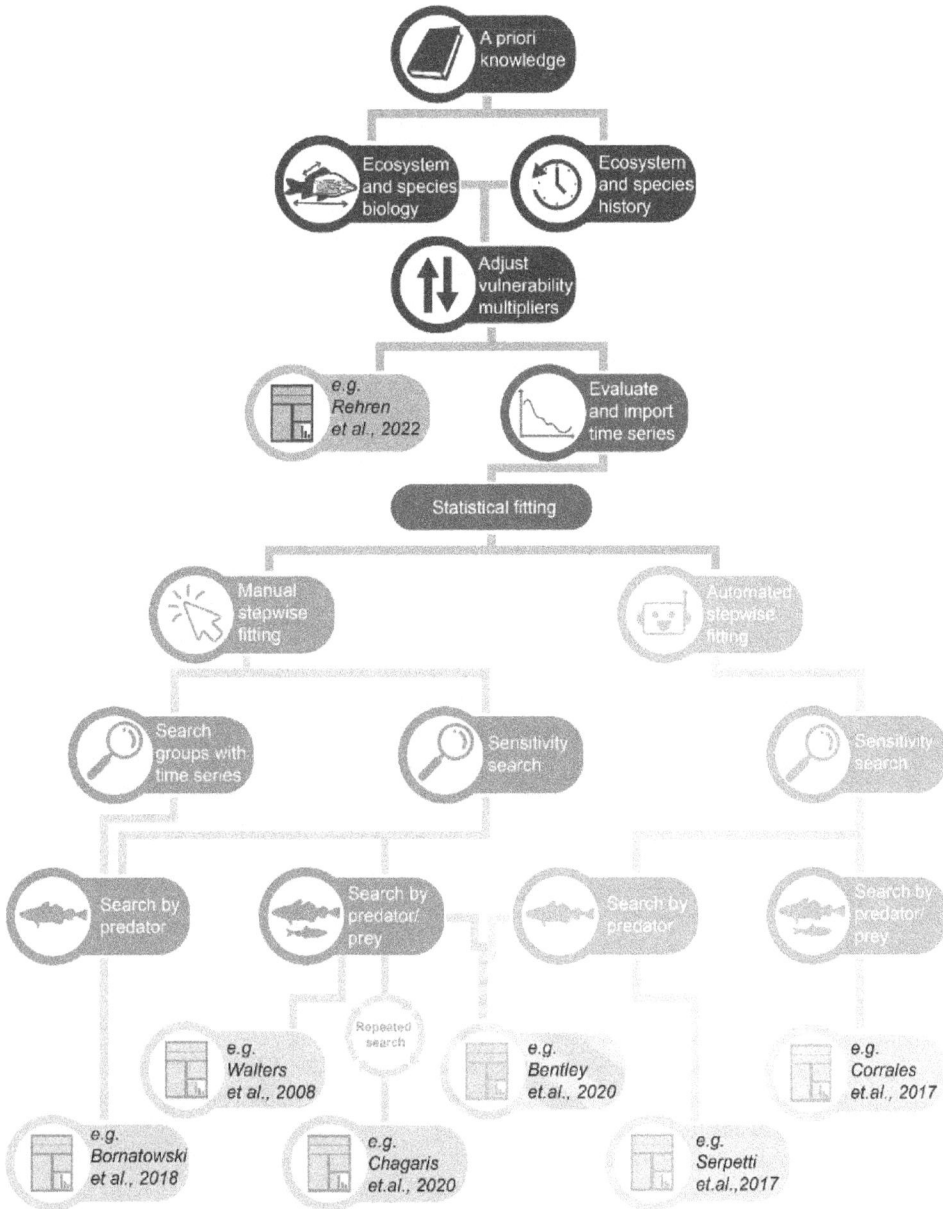

Figure 2. Pathways for estimating vulnerability multipliers (k_{ij}) in Ecosim. All pathways end with a reference to peer-reviewed examples.

Rather than citing this chapter, please cite the source.

Media Attributions

- From Bentley et al. 2024 Figure 1
- From Bentley et al. 2024. Figure 2

Notes

1. Walters, C.J., Martell, S.J. 2004. Fisheries ecology and management, Vol., Princeton University Press, Princeton, New Jersey

2. Walters, C., Christensen, V., Pauly, D. 1997. Structuring dynamic models of exploited ecosystems from trophic mass-balance assessments. Reviews in Fish Biology and Fisheries 7: 139-172. https://doi.org/10.1023/A:1018479526149

3. Walters, C., Pauly, D., Christensen, V. and Kitchell, J.F., 2000. Representing density dependent consequences of life history strategies in aquatic ecosystems: EcoSim II. Ecosystems, 3(1): 70-83. https://doi.org/10.1007/s100210000011

4. Ahrens, R.N.M., Walters, C.J. and Christensen, V. (2012), Foraging arena theory. Fish and Fisheries, 13: 41-59. https://doi.org/10.1111/j.1467-2979.2011.00432.x

5. Walters et al., 1997. *op. cit.*

6. e.g., Rehren, J., Coll, M., Jiddawi, N., Kluger, L.C., Omar, O., Christensen, V., Pennino, M.G. and Wolff, M., 2022. Evaluating ecosystem impacts of gear regulations in a data-limited fishery—comparing approaches to estimate predator–prey interactions in Ecopath with Ecosim. ICES Journal of Marine Science 79(5):1624-1636. https://doi.org/10.1093/icesjms/fsac077

7. e.g., Scott, E., Serpetti, N., Steenbeek, J. and Heymans, J.J., 2016. A Stepwise Fitting Procedure for automated fitting of Ecopath with Ecosim models. SoftwareX, 5, pp.25-30. https://doi.org/10.1016/j.softx.2016.02.002

30.

Using ecology and history to derive vulnerability multipliers

Jacob Bentley; David Chagaris; Marta Coll; Sheila JJ Heymans; Natalia Serpetti; Carl J. Walters; and Villy Christensen

Vulnerability multipliers are perhaps easiest to understand when it is recognized that they reflect how far an exploited predator is from its carrying capacity (e.g., interpreted as unfished state); vulnerability multipliers should allow consumption rates that enable a species to recover from its Ecopath biomass to its unfished biomass if fishing ceases. EwE can use the ratio between a group's unfished biomass and its Ecopath base biomass to estimate vulnerability multipliers for exploited groups, see, e.g., the EwE User Guide vulnerability multiplier estimator chapter.

An added corollary is that the unfished state may be associated with high abundance of top predators and low abundance of their prey due to high predation mortality. If such top predator populations are fished down, predator release may cause the prey to increase. For those prey, the vulnerability multipliers should thus be set to a high value, even though the baseline model represents the unfished state.

It can indeed be difficult to specify reasonable vulnerability multipliers for non-exploited species. Here, vulnerability multipliers need to be considered in the context of the foraging arena: the fine-scale spatial structure of the trophic interactions and what proportion of prey may be vulnerable to predation at any moment (Figure 1). The activity, spatial restrictions, and distributions of species provide insight into the likely vulnerability of prey to predation. This in turn provides a starting point from which it is possible to assign vulnerability multipliers. The distribution of predators could be restricted by limited mobility, habitat requirements, or the predation risk they face themselves, whereas prey vulnerability may be influenced by the time they spend in and out of safe behavioral states. This can be related to the availability of shelter, such as macroalgae, or specific ontogenetic life stages (for example juvenile fish may allocate less time to foraging), or be more restricted spatially (and thus unable to access vulnerable pools of prey) than their adult counterparts. Different behaviours, such as dispersal behaviours (e.g., moving to spawning sites), aggressive behaviours, or evolutionary behaviour (e.g., changes in shoaling dynamics) may also influence vulnerability to predation.

Trophic levels have also been used to approximate vulnerability multipliers in situations where time series data were unavailable under the dubious assumption that the vulnerability multipliers are proportional to the trophic level of the predator. This approach assumes that higher trophic levels are further removed from their unfished biomass than lower trophic levels, typically because of historical overfishing. This may seem a reasonable assumption considering how global fisheries have historically targeted and depleted higher trophic level fish stocks[1 2], but conflicts with the concept of using *a priori* knowledge to parameterize vulnerability multipliers based on region specific trends in historical exploitation or ecology.

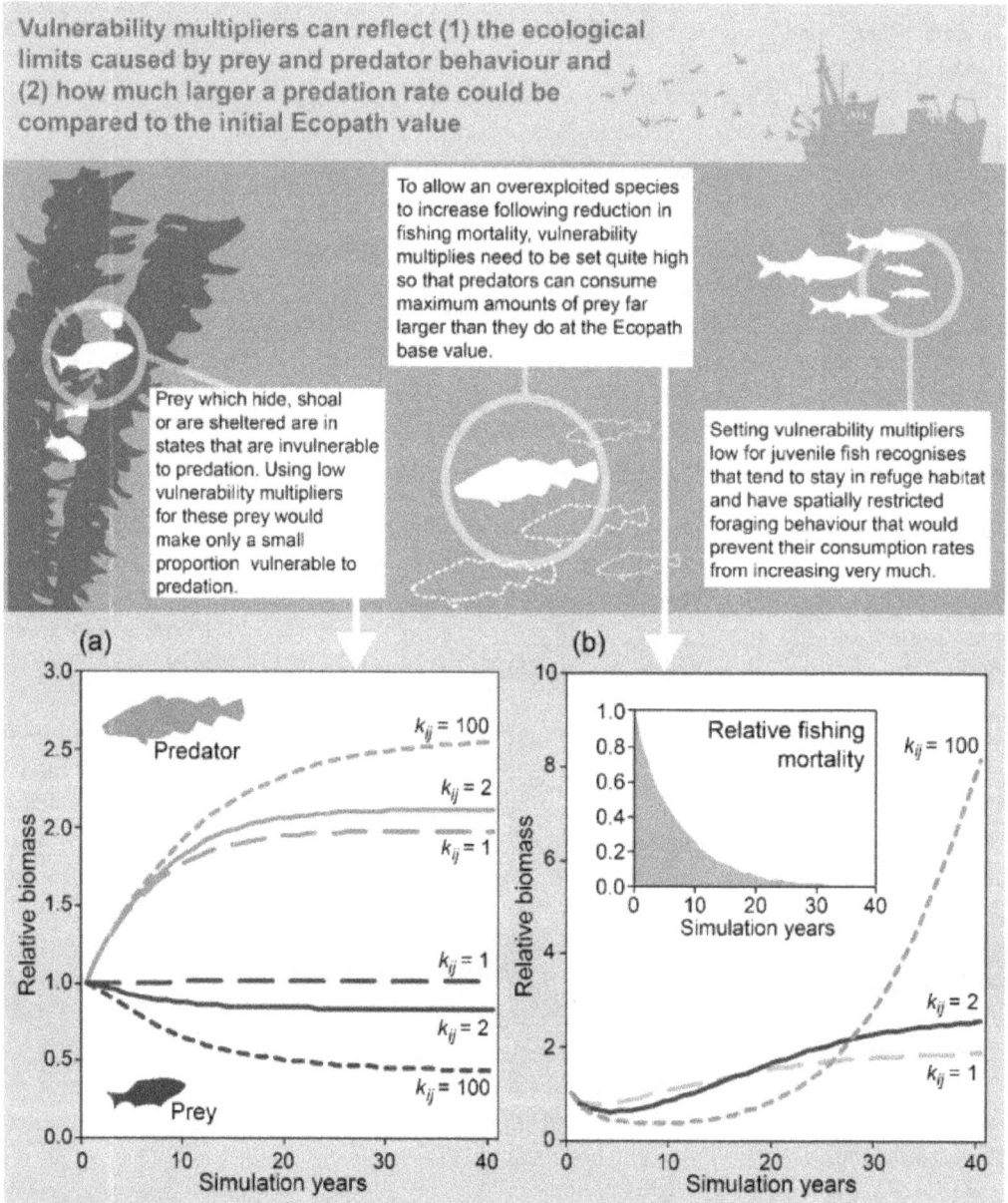

Vulnerability multipliers can reflect (1) the ecological limits caused by prey and predator behaviour and (2) how much larger a predation rate could be compared to the initial Ecopath value

To allow an overexploited species to increase following reduction in fishing mortality, vulnerability multiplies need to be set quite high so that predators can consume maximum amounts of prey far larger than they do at the Ecopath base value.

Prey which hide, shoal or are sheltered are in states that are invulnerable to predation. Using low vulnerability multipliers for these prey would make only a small proportion vulnerable to predation.

Setting vulnerability multipliers low for juvenile fish recognises that tend to stay in refuge habitat and have spatially restricted foraging behaviour that would prevent their consumption rates from increasing very much.

Figure 1. Using history and ecology to estimate vulnerability multipliers (k_{ij}) in Ecosim. The illustration at the top of the figure provides examples where k_{ij} estimates can be inferred from functional group ecology and life history. Model simulations demonstrate how (a) prey vulnerability influences predator-prey biomass trajectories when predator biomass increases and (b) how k_{ij} estimates impact the rate of functional group recovery following reduction in fishing (inset figure).

Finally, an approach to setting vulnerability multipliers was applied by Chagaris et al.[3] to constrain how much predation mortality by a given predator could increase relative to a prey's total natural mortality

$$k_{ij} = \frac{M2_{cap} \cdot M_i}{M2_{base,ij}} \tag{1}$$

where $M2_{cap}$ defines the proportion of the natural mortality of a prey that a predator can be responsible for, M_i is the natural mortality of prey i, and $M2_{base.ij}$ is the base predation mortality by predator j on prey i. There may be ecological reasons, or reasons derived from data, to prevent a single predator from being accountable for large proportions of a prey's natural mortality. Using these k_{ij} instead of default or model estimated values, (which are often higher) may also be driven by ambitions for model performance: Chagaris et al.[4] found that extremely high k_{ij} estimated by Ecosim led to instability at high fishing mortality rates when evaluating equilibrium yield curves, and using $M2_{cap}$ values between 0.75 to 1.0 led to more reasonable estimates for F_{MSY} (the fishing mortality at maximum sustainable yield) while also constraining theoretical maximum predation mortality rates to values that were compatible with prey natural mortality rates.

Attribution
This chapter is based on Bentley JW, Chagaris D, Coll M, Heymans JJ, Serpetti N, Walters CJ and Christensen V. 2024. Calibrating ecosystem models to support marine Ecosystem-based Management. ICES Journal of Marine Science, https://doi.org/ 10.1093/icesjms/fsad213. Adapted based on CC BY License.
Rather than citing this chapter, please cite the source.

Media Attributions

- From Bentley et al. 2024. Figure 3

Notes

1. Christensen, V. 1996. Managing fisheries involving top predator and prey species components. Reviews in Fish Biology and Fisheries. 6:417-442.

2. Pauly, D., V. Christensen, A. Dalsgaard, R. Froese, and J. Torres. 1998. Fishing down marine food webs. Science 279 (5352): 860-863. DOI: 10.1126/science.279.5352.860

3. Chagaris, D., Drew, K., Schueller, A., Cieri, M., Brito, J. and Buchheister, A., 2020. Ecological reference points for Atlantic menhaden established using an ecosystem model of intermediate complexity. Frontiers in Marine Science, 7, p.606417. https://doi.org/10.3389/ fmars.2020.606417

4. Chagaris et al. 2020. *op. cit.*

31.

Statistical approaches for estimating vulnerability multipliers

Jacob Bentley; David Chagaris; Marta Coll; Sheila JJ Heymans; Natalia Serpetti; Carl J. Walters; and Villy Christensen

It is becoming more common for Ecosim vulnerability multipliers and primary production anomalies to be estimated using statistical routines, with Heymans et al.[1] demonstrating that it is best practice to estimate vulnerability multipliers by fitting model simulations to time series reference data. Longer time series are preferable as they provide an opportunity to explore important drivers of change and tend to have strong contrast in the data, which improves the model's ability to estimate parameters, leading to more confidence in our assessment of ecosystem dynamics.

Model fitting that includes estimation of primary production anomalies is basically the ecosystem equivalent of estimating recruitment and mortality anomalies in state-space approaches to parameter estimation for single-species models, a key difference being that the anomaly estimates may be informed by correlated variation in time series patterns of multiple species.

The quality (precision, informative contrast over time) of time series data is important, especially if the fitted model is to be used for management purposes, as vulnerability multipliers (and thus predation rate changes), will be used in forward simulations to times beyond the data. In Ecosim, users can weight time series data to represent how reliable or variable time series are compared to the other reference time series. Low weights imply that the data either has high variance or is unreliable (e.g., underestimated or uncertain catches). Weightings impact the contribution of time series to the assessment of model performance, where a weight of 0 indicates that the time series will not be used in the calculation of "goodness of fit". Weightings can be assigned based on a qualitative assessment of data pedigree (e.g., based on data origin), or by using more quantitative information, such as confidence intervals from survey estimates, the retrospective analyses of stock assessment models, or signal to noise ratio assessments[2].

The procedure for estimating vulnerability multipliers and production anomalies that improve the fit of model simulation to calibration data is based on minimization of a sum of squares (*SS*) of prediction errors, which is then checked for overparameterization using the Akaike Information Index *(AIC)*[3] [4].

The *SS* is used to calculate a log likelihood criterion (Figure 1), assuming normally distributed deviations of log model predictions from log observations, evaluated at the conditional maximum likelihood estimate of the prediction error variance and scaled in the case of relative observations *(y)* by the maximum likelihood estimate of the relative simulation scaling factor *(q)* in the equation $y = q \times X$, where X is the absolute observation. Model fitting then proceeds by numerical search procedures to seek parameter values that minimize SS.

When generating a set of model fits under different fitting hypotheses or methods for choosing what parameters to include in the SS, *AIC* is then used to identify the model of best fit. *AIC* is a tool for model selection that penalizes for fitting too many parameters relative to the time series available for estimating the SS and is calculated as

$$AIC = n \cdot \ln(\frac{minSS}{n}) + 2K \tag{1}$$

where *n* is the total number of observations, or time series values, from the loaded calibration time series and *K* is the number of parameters estimated. When sample size is small, there is a large probability that *AIC* will select models with too many estimated parameters (i.e., overfit models). The modified AIC_c can be used to address this potential overfitting by including a correction for small sample sizes

$$AIC_c = AIC + 2K \cdot (\frac{K-1}{n-K-1}) \tag{2}$$

As a rule of thumb[5], AIC_c should be used unless $n/K > {\sim}40$. In other words, unless the number of estimated parameters equates to a minimum 2-3% of the amount of data, use AIC_c. AIC_c should therefore be used when assessing EwE model performance.

A word of caution, the AIC calculations assume that the observations are independent whereas timeseries data such as typically used for ecosystem modelling have high autocorrelation. For this reason, it is advisable to test the impact on assumptions about *n* on model selection.

Formal estimation

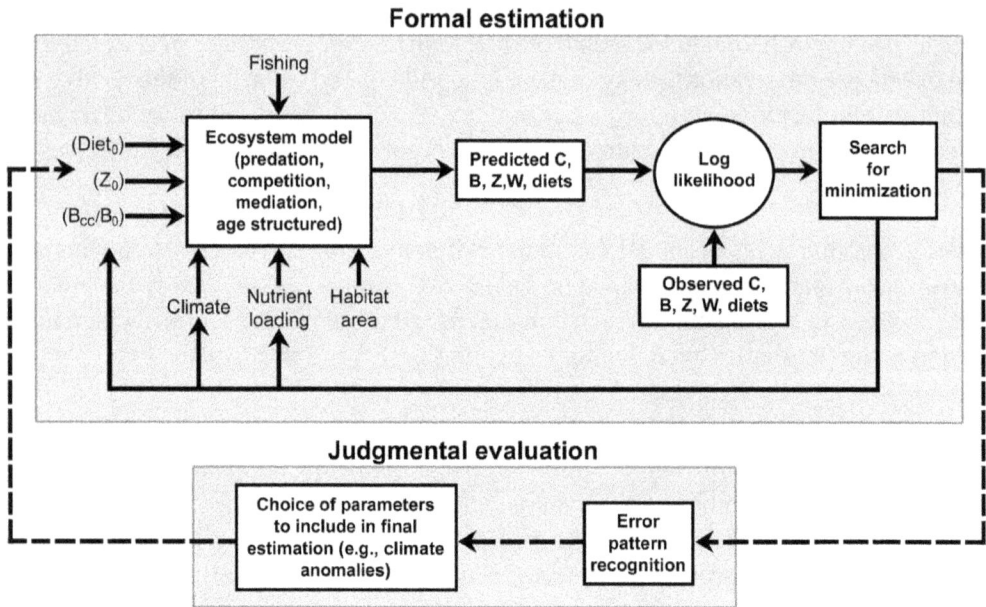

Figure 1. **Overview of the Ecopath and Ecosim modelling process. Using log likelihood criteria, vulnerability multipliers or production anomalies (e.g., climate or nutrient loading) may be estimated based on a non-linear search routine and vulnerability multiplier (vulmult) estimation. Prediction (fitting) failures after each estimation trial then inform judgmental changes in model structure and parameters. B is biomass, Z is total mortality, C is catch, W is average weight. Subscript 0 refers to the Ecopath model base year, and CC to carrying capacity. B_{cc}/B_0 refers to vulmult. From Christensen and Walters[6].**

Multiple approaches have been developed to statistically estimate vulnerability multipliers (see Figure 2 in Vulnerability and vulnerability multipliers chapter). They can be estimated for predators, providing a single multiplier limit to all of a given predator's base predation rates, and they can be estimated for individual predator-prey relationships, which assumes that the multiplier limits are heterogeneous across prey. This choice tends to be associated with user preference, ecological justification, or determined based on the approach that produces the best fit model. Whether estimating predator or predator-prey vulnerability multipliers, there are a few ways to select which vulnerability multipliers should be estimated:

1. manual selection based on a priori knowledge or species priority;

2. select vulnerability multipliers for groups with calibration time series; or

3. select the most sensitive vulnerability multipliers (i.e., those that when changed have the largest impact on *SS*).

Manually selecting vulnerability multipliers allows for an early integration of ecological information but may lead to a sub-optimal model fit if the *SS* is not sensitive to the selected multipliers. Conversely, the sensitivity search may optimise model fit but it is purely statistical and does not know what makes sense ecologically. Only estimating vulnerability multipliers for groups with time series acknowledges that, to some degree, the parameter should be constrained by the available time series. The level of group connectedness within the food web (e.g., group consumption and predation) may also constrain the parameter search if changes in vulnerability multipliers impact the contribution of other groups. Groups that do not have informative time series or, have low connectedness in the food web, have widely variable estimated vulnerability multipliers – search routines can change those without any penalty incurred.

Choosing how many vulnerability multipliers to estimate, without overfitting is another point of confusion and discussion. The number of vulnerability multipliers that can be potentially estimated is often significantly more than the data available to constrain simulations. EwE best practices suggest that a conservative number of Degrees of Freedom (DoF) and therefore parameters to estimate is one less than the number of calibration time series available[7]. This approach recognizes that values within time series are highly autocorrelated, viewing each time series as an "independent observation", but the approach could be overly conservative, especially if long time series are available– especially if the contrast (ups and downs) and are not just one-way trajectories.

Both manual and automated statistical calibration routines are available in Ecosim to search for vulnerability multipliers. The manual approach can be arduous when testing multiple fitting hypotheses (e.g., with or without fishing effort or primary production anomalies) as the number of plausible fitting combinations can easily reach the hundreds, if not thousands, increasing the likelihood of user error. In the past, users have overcome this issue by only testing the nth fitting scenario (e.g., 5, 10, 15 vulnerability multipliers etc.)[8] However this approach risks overlooking the vulnerability multiplier combination, which produces the best statistical fit. The stepwise fitting procedure developed by Scott et al.[9] automates this process, allowing for a broad exploration of the parameter space which accelerates the process and removes the problem of user error. Recent improvements to the automated approach have increased the computational speed by enabling multiple fitting scenarios to be tested simultaneously using computers multithreading capabilities (J. Steenbeek, pers. comm.).

Novel approaches to estimate vulnerability multipliers using the manual and automated fitting routines have also been developed for two EwE models which are being used operationally to inform fisheries catch advice. Bentley et al.,[10] employed an approach, which combined searches for predator vulnerability multipliers and predator-prey vulnerability multipliers, whereas the approach developed by Chagaris et al.[11] uses the manual fitting tool in Ecosim to iteratively estimate the most sensitive predator-prey vulnerability multipliers over multiple sequential (repeated) tuning iterations.[12].

It is worth reiterating that statistical estimation of vulnerability multipliers does not necessarily have any bearing on ecology. While it is possible to exclude vulnerability multipliers from the search routine, there is currently no mechanism to include prior information or ecologically sensible bounds to constrain the limits for vulnerability mul-

tipliers included in the search routine. A judgement evaluation following the formal estimation of vulnerability multipliers should be applied to:

1. reflect on the ecological assumptions attached to estimated vulnerability multipliers,

2. assess how realistic functional group simulations are (in hindcast and future),
   and

3. understand and fix issues with model structure and parameterization (Figure
   1).

It is possible to view the fit of each functional group to calibration time series and its
contribution to the overall *SS* in Ecosim via the *Ecosim > Output > Ecosim group plots*
form. This is often used to screen issues with model simulations, such as contradicting
trends or misalignment in initial time steps, and direct fixes.

However, what is often not accounted for when estimating vulnerability multipliers is
their impacts on the advice products such as F_{MSY} or food web indicators. The focus is
often only on the goodness of fit of the model, but the impacts of estimated vulnerability
multipliers on predictions and reference points should be evaluated[13]. We next provide
two case studies to explore how alternate fitting approaches impact the emergence of vulnerability multipliers and how vulnerability multipliers impact model outputs.

Attribution
This chapter is based on Bentley JW, Chagaris D, Coll M, Heymans JJ, Serpetti N,
Walters CJ and Christensen V. 2024. Calibrating ecosystem models to support marine
Ecosystem-based Management. ICES Journal of Marine Science, https://doi.org/
10.1093/icesjms/fsad213. Adapted based on CC BY License. Rather than citing this
chapter, please cite the source.

Media Attributions

- Christensen and Walters, 2011, Figure 1.

Notes

1. Heymans, J.J., Coll, M., Link, J.S., Mackinson, S., Steenbeek, J., Walters, C. and Christensen,
   V., 2016. Best practice in Ecopath with Ecosim food-web models for ecosystem-based management. Ecological Modelling, 331, pp.173-184. https://doi.org/10.1016/
   j.ecolmodel.2015.12.007

2. Heymans et al., 2016. *op. cit.*

3. Akaike, H., 1998. Information theory and an extension of the maximum likelihood principle.
   In Selected papers of Hirotugu Akaike (pp. 199-213). Springer, New York, NY.

https://doi.org/10.1007/978-1-4612-1694-0_15

4. Cavanaugh, J.E. and Neath, A.A., 2019. The Akaike information criterion: Background, derivation, properties, application, interpretation, and refinements. Wiley Interdisciplinary Reviews: Computational Statistics, 11(3), p.e1460. https://doi.org/10.1002/wics.1460

5. Burnham, K. P., & Anderson, D. R. (2004). Multimodel Inference: Understanding AIC and BIC in Model Selection. Sociological Methods & Research, 33(2), 261–304. https://doi.org/10.1177/0049124104268644

6. Christensen and Walters. 2011. Op. cit.

7. Heymans et al., 2016. *op. cit.*

8. e.g., Alexander, K.A., Heymans, J.J., Magill, S., Tomczak, M.T., Holmes, S.J. and Wilding, T.A., 2015. Investigating the recent decline in gadoid stocks in the west of Scotland shelf ecosystem using a foodweb model. ICES Journal of Marine Science, 72(2), pp.436-449. https://doi.org/10.1093/icesjms/fsu149

9. Scott, E., Serpetti, N., Steenbeek, J. and Heymans, J.J., 2016. A Stepwise Fitting Procedure for automated fitting of Ecopath with Ecosim models. SoftwareX, 5, pp.25-30. https://doi.org/10.1016/j.softx.2016.02.002

10. Bentley, J.W., Serpetti, N., Fox, C.J., Heymans, J.J. and Reid, D.G., 2020. Retrospective analysis of the influence of environmental drivers on commercial stocks and fishing opportunities in the Irish Sea. Fisheries Oceanography, 29(5), pp.415-435. https://doi.org/10.1111/fog.12486

11. Chagaris, D., Drew, K., Schueller, A., Cieri, M., Brito, J. and Buchheister, A., 2020. Ecological reference points for Atlantic menhaden established using an ecosystem model of intermediate complexity. Frontiers in Marine Science, 7, p.606417. https://doi.org/10.3389/fmars.2020.606417

12. Full methodologies for these approaches are provided in the Bentley et al. 2024, Supplementary Material.

13. e.g., Rehren, J., Coll, M., Jiddawi, N., Kluger, L.C., Omar, O., Christensen, V., Pennino, M.G. and Wolff, M., 2022. Evaluating ecosystem impacts of gear regulations in a data-limited fishery—comparing approaches to estimate predator–prey interactions in Ecopath with Ecosim. ICES Journal of Marine Science 79(5):1624-1636. https://doi.org/10.1093/icesjms/fsac077

32.

Case study: Fitting impact on vulnerability multipliers

Jacob Bentley; David Chagaris; Marta Coll; Sheila JJ Heymans; Natalia Serpetti; Carl J. Walters; and Villy Christensen

The Anchovy Bay ecosystem model that is used to describe and test EwE scenarios throughout this text book was used to investigate how vulnerability multipliers emerge (and whether they re-emerge through fitting) and how this process is influenced by:

1. noise in the calibration data, and

2. the chosen approach for estimating vulnerability multipliers: "predator" or "predator-prey" vulnerability multipliers.

We investigated the impact of emerging vulnerability multipliers on biomass and catch simulations and estimates of fishing mortality consistent with maximum sustainable yield (F_{MSY}).

Building a base Ecosim model[1]

Ecosim simulations for Anchovy Bay were created by adding temporal trends to fishing effort and adjusting vulnerability multipliers. Simulated fishing effort trends reflected trends often seen in reality:

- sealers fishing effort followed an exponential decline as may be expected in response to conservation efforts/policy,

- trawlers fishing effort followed an exponential decline under the assumption that whitefish (cod and whiting) stocks have been overexploited, leading to reductions in effort to encourage stock recovery,

- seiners and bait boat effort followed a slight linear increase in response to growing demand, and

- shrimpers effort increased assuming fishers shifted their target species to shrimp following reduced opportunities to catch white fish.

Vulnerability multipliers (k_{ij}) were adjusted following ecological assumptions and assumptions linked to the fishing effort trajectories. To distinguish between scenarios more easily, predator vulnerability multipliers will hereafter be denoted as k_j, while predator-prey vulnerability will remain as k_{ij}. For predator estimates, a mix of high, low and default estimates were applied. For groups which were assumed to be overexploited, values were estimated using the "Estimate Vulnerabilities" interface. For the predator-prey estimates, the Ecosim sensitivity search was used to identify the 10 most sensitive predator/prey k_{ij} parameters, which were then adjusted to ensure a range of high and low k_{ij} estimates were included.

For the purpose of this exercise, these two simulations (one with predator k_j and one with predator-prey k_{ij}) were viewed as perfect representations of their ecosystems, i.e., the biomass and catch simulations were "real observations" driven by the "true" vulnerability multipliers. The aim of the following exercise was to test whether, when using these "real observations" as calibration time series, the "true" vulnerability multipliers would reemerge, and whether the addition of noise to the "real observations" had any impact on the emerging vulnerability multipliers. Biomass and catch simulations were extracted from Ecosim and four scenarios for observation data quality were prepared: noise (random noise, normally distributed around the mean (true) biomass trend to represent observation error) was added to the calibration time series with coefficients of variation (CV) of 0 (no noise) 0.1, 0.3, and 0.5.

Predator vulnerability multipliers

Vulnerability multipliers were reset to the default value of 2; fishing dynamics were not changed from those used to produce the "real observations." The exported biomass and catch time series were used as calibration time series to estimate predator vulnerability multipliers for the functional groups seals, cod, whiting, shrimp, benthos, and zooplankton using the manual stepwise fitting interface. k_j values for groups which had values of 2 in the initial model were not altered.

Figure 1 shows how k_j parameters emerged after model calibration, and how this altered functional group carrying capacities in the absence of fishing and F_{MSY} estimates. k_j values which emerged when estimated using the calibration time series with no noise were similar to the "true" parameters (Figure 1a). Adding noise to the calibration time series led to divergence between the estimated k_j values and the "true" parameters, highlighting the impact data quality can have on the fitting procedure and thus stressing the importance of evaluating the suitability of time series before using them to drive model calibration. The variability in k_j re-emergence under the four data quality scenarios was also unique to specific functional groups, for example: k_j estimates for cod showed greater re-emergence accuracy (or consistency) when compared to other functional groups. Cod is highly connected within the food web (i.e., cod is an opportunistic predator which is also preyed upon by higher trophic levels), therefore vulnerability multipliers which improve the model fit tend to be more constrained due to their potential to have large cascading impacts on the wider food web. In addition, cod also experienced a period of collapse followed by recovery, which provides much needed contrast for the model to reliably estimate the vulnerability multipliers.

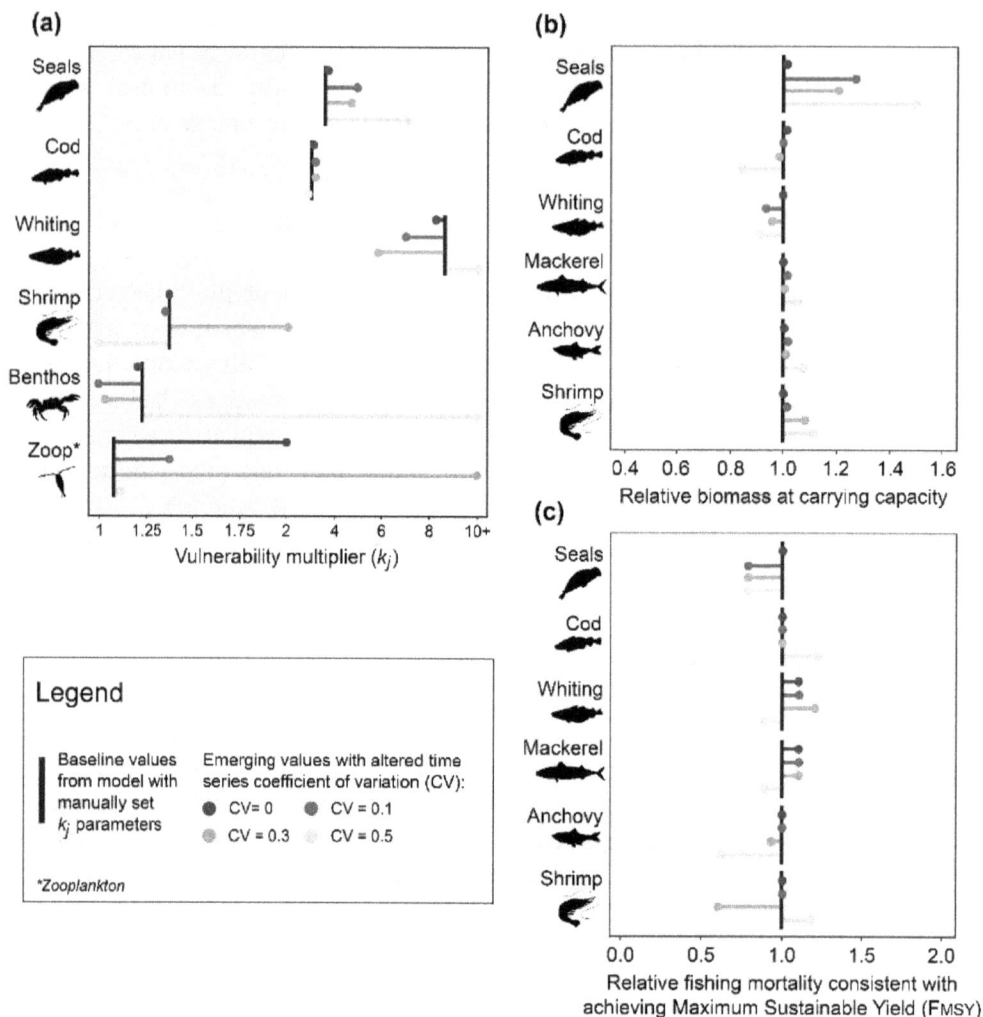

Figure 1. Estimation and impact of predator vulnerability multipliers (k_j). The Anchovy Bay ecosystem model was calibrated against generated time series with incremental coefficients of variation (*CV*) to identify the impact of time series quality on (a) k_j re-emergence and how k_j estimates impacted (b) functional group carrying capacities in the absence of fishing and (c) estimates of relative fishing mortality consistent with achieving Maximum Sustainable Yield (F_{MSY}).

Functional group carrying capacities and estimates of F_{MSY} were impacted by emerging values (Figure 1b and 1c). Carrying capacities from scenarios with parameters calibrated against data with no noise were most similar to those achieved with the "true" parameters (Figure 1b), with dissimilarity generally increasing with the addition of noise to the calibration data. The importance of acknowledging the impact of estimates beyond model fit is demonstrated with the resulting F_{MSY} estimates: relative changes to F_{MSY} estimates mirrored the deviations of estimated values relative to the "true" values (Figure 1c). Increases in values led to decreases in F_{MSY}, while decreases in values led to increases in F_{MSY} This is because higher values enable groups to recover faster with the cessation of fishing and reach a higher carrying capacity, but they also decrease stock resilience to increases in F (functional groups decline faster and more severely if you increase their). It is worth noting that where differences between true and estimated F_{MSY} occurred, they were not proportional to the difference in true and estimated vulnerability multipliers (i.e., large changes in k_{ij} do not result in equally large changes to F_{MSY}).

Predator-prey vulnerability multipliers

Similar to the predator scenario, vulnerability multipliers were reset to the default of 2, and the exported biomass and catch time series (generated with "true" predator-prey vulnerability multipliers) were used as calibration time series to estimate predator-prey vulnerability multipliers. Predator-prey values for the ten most sensitive predator/prey parameters were estimated using the manual stepwise fitting interface. Figure 2 shows how parameters emerged and how this altered functional group carrying capacities and FMSY estimates.

In comparison to the emergence of predator vulnerabilities, the emergence of predator-prey vulnerabilities was less constrained with examples of poor re-emergence accuracy across all calibration data scenarios (Figure 2a). Functional group carrying capacities showed higher dissimilarity from their baseline when compared to predator simulations and their baseline (Figure 2b). Carrying capacity dissimilarity increased with the addition of noise to the calibration data, however simulations with no/low noise were notably more dissimilar when estimating predator-prey vulnerabilities as opposed to predator vulnerabilities (Figure 2b) which is due to the greater differences in k_{ij} estimates.

Relative F_{MSY} estimates, influenced by predator-prey k_{ij} values, showed higher dissimilarity from their baseline (Figure 2C) when compared to F_{MSY} estimates influenced by predator k_j values (Figure 1C). The links between predator-prey k_{ij} values and F_{MSY} are less obvious than the links between predator k_j values and F_{MSY} due to the more complex interaction-specific consumption limits. This is particularly true for groups with mixed diets (e.g., cod, whiting, seals, and mackerel) while links between predator-prey values and the F_{MSY} estimates for groups, which are heavily dependent on a single prey group were observed for anchovy (F_{MSY} mirrors the anchovy/zooplankton k_{ij} estimates) and shrimp (F_{MSY} mirrors the shrimp/benthos k_{ij} estimates).

(a)

(b)

(c)

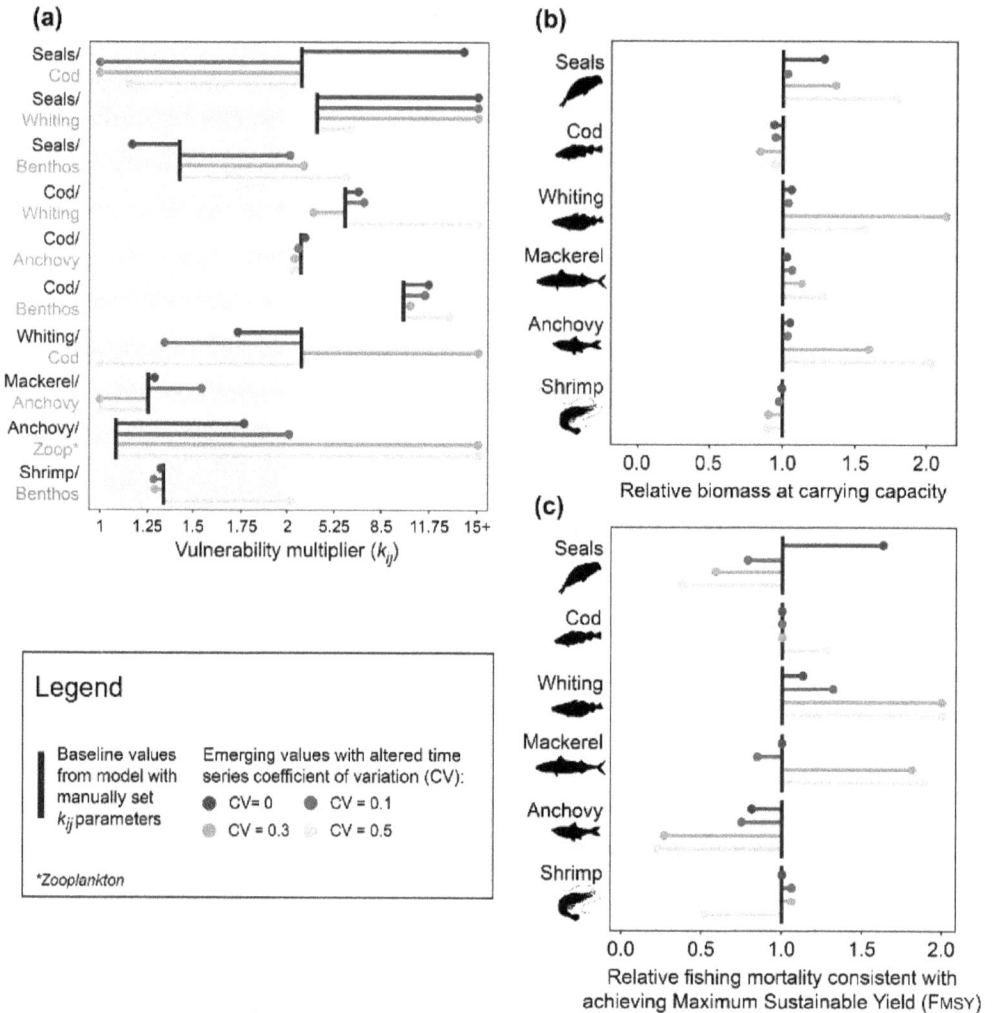

Figure 2. Estimation and impact of predator-prey vulnerability multipliers (k_{ij}). The Anchovy Bay ecosystem model was calibrated against generated time series with incremental coefficients of variation (*CV*) to identify the impact of time series quality on (a) k_{ij} re-emergence and how k_{ij} estimates impacted (b) functional group carrying capacities in the absence of fishing and (c) estimates of relative fishing mortality consistent with achieving Maximum Sustainable Yield (F_{MSY}).

Media Attributions

- From Bentley et al. 2024 Figure 5
- From Bentley et al. 2024. Figure 6

Notes

1. See Bentley et al. 2024 Supplementary Data for details about the model construction

33.

Case study: How fitting impacts advice

Jacob Bentley; David Chagaris; Marta Coll; Sheila JJ Heymans; Natalia Serpetti; Carl J. Walters; and Villy Christensen

The Irish Sea EwE model[1] and Northwest Atlantic Continental Shelf EwE model[2] (hereafter called NWACS-MICE) have both been used to inform fisheries advice for their respective regions using ecological/ecosystem reference points[3]. Both models were designed to focus on commercial fisheries however, they have very different structures in terms of model complexity (Table 1). The two models were used to demonstrate the outcomes and management implications of vulnerability multiplier (k_j or k_{ij}) estimation and compared estimates of F_{MSY} and ecosystem indicators. Ecosystem indicators selected for this analysis included total system biomass, commercial biomass, total catch, system diversity (Kempton's Q), the trophic level of the catch, and the trophic level of the community. Estimates of F_{MSY} and ecosystem indicators were compared across the following nine fitting approaches:

1. Predator k_j values, where the number of parameters estimated is one less than the number of available calibration time series *(K-1)*.

2. Predator k_j values estimated for all functional groups with time series.

3. Predator k_j values using the automated stepwise fitting approach, where the applied k_j values are taken from the model with the lowest AIC_c, and the vulnerabilities are reset to the default (2) at each fitting iteration.

4. Predator k_j values using the automated stepwise fitting approach, where the applied k_j values are taken from the model with the lowest AIC_c, and the vulnerabilities are retained from previous fitting iterations.

5. Predator-prey k_{ij} values, where the number of parameters estimated is one less than the number of available calibration time series *(K-1)*.

6. Predator-prey k_{ij} values using the automated stepwise fitting approach, where the applied values are taken from the model with the lowest AIC_c, and the vulnerabilities are reset to the default (2) at each fitting iteration.

7. Predator-prey k_{ij} values using the automated stepwise fitting approach, where the applied values are taken from the model with the lowest AIC_c, and the vulnerabilities are retained from previous fitting iterations.

8. Predator-prey k_{ij} values using a repeated manual stepwise fitting approach, where the estimated k_{ij} *(K-1)* are retained from one iteration to the next (with a

total of 5 iterations) and the final configuration is that with the lowest AIC_c, as was done in Chagaris et al.,[4].

9. A combination of predator k_j and predator-prey k_{ij} values using the methods outlined by Bentley et al., (2020). Predator k_j values were estimated using the automated stepwise fitting approach in #3. Predator-prey k_{ij} values were esti-mated using a manual stepwise fitting approach and the remaining degrees of freedom. The number of additional predator-prey k_{ij}'s was determined by their AIC_c score (note this approach was only carried out for the Irish Sea model in this study).

Table 1. Comparison of key characteristics between static (Ecopath) and time-dynamic (Ecosim) models of the Irish Sea and Northwest Atlantic Continental Shelf (NWACS-MICE).

Static model	Irish Sea EwE model	NWACS-MICE
Total number of functional groups	41	17
Number of multi-stanza groups	4	5
Fleets	8 (gear specific)	8 (species specific?)
Time dynamic model	Irish Sea EwE model	NWACS-MICE
Time-period	1973-2016	1985-2017
Calibration time series	52 (28 biomass, 24 catch)	28 (18 biomass, 10 catch)
Fitting approach	Predator and predator-prey search	Repeated predator-prey search

Alternate fitting approaches led to the emergence of different vulnerability multipliers in the corresponding models of best fit (as determined by sum of squared deviations and AIC_c) for the Irish Sea (Figure 1) and NWACS-MICE (Figure 2). Different fitting approaches impacted estimates of F_{MSY} in both models due to changes in species sensi-tivity to F with alternate vulnerability multipliers. Despite the increased complexity of the Irish Sea model, the patterns in F_{MSY} variability are comparable between models, with certain species having consistent F_{MSY} estimates across the nine approaches for vulner-ability multiplier estimation. This includes cod and whiting for the Irish Sea and striped bass for the NWACS. As demonstrated in case study 1 (previous chapter), these species are opportunistic feeders which are also predated on by higher trophic levels, giving them a relatively high degree of connectivity within the food web models, which may con-strict the emergence of vulnerability multipliers. Additionally, these groups have experi-enced a period of collapse, and in some cases recovery, which provides contrast for the model to estimate the vulnerability multipliers. Fitting approaches with similar properties resulted in more closely related F_{MSY} estimates[5]. For example, F_{MSY} estimates gener-ated by approaches, which searched for vulnerability multipliers by "predator" tended to be more similar to each other when compared to those generated by approaches which searched by "predator-prey", and *vice versa*. This emergent trend is perhaps most clearly seen in the F_{MSY} estimates for menhaden and bluefish adults from the NWACS-MICE model (Figure 2).

The approach used to estimate vulnerability multipliers had an impact on the derived ecosystem indicators (Figure 1b and Figure 2b). These impacts were relatively small, most deviations being within the range of 5-10% when compared to indicators from the published models. Trophic level indicators were particularly robust across estimation approaches, despite often larger differences being observed in diversity (Kempton's Q) catch and commercial biomass indicators. Balanced reconfiguration within the ecosystem models (i.e., increases in some species and decreases in others with similar trophic levels) enabled the trophic indicators to remain similar across approaches. However, the dissimilarity in trophic level of the catch in the NWACS-MICE model was generally higher across scenarios where vulnerability multipliers were searched by "predator". This reflects the higher F_{MSY} reference points for adult weakfish and bluefish and lower FMSY reference points for menhaden and herring produced under the same fitting approaches. Overall, the Irish Sea EwE model showed greater dissimilarity in derived indicators than the NWACS-MICE model. This outcome is likely linked to the increased complexity of the Irish Sea model, and how a repeated search provides the opportunity to adjust more predator-prey vulnerability multipliers. This may be less of a concern for low complexity models as the parameter space is smaller, increasing the likelihood that the same vulnerability multipliers will be adjusted.

(a)

(b)

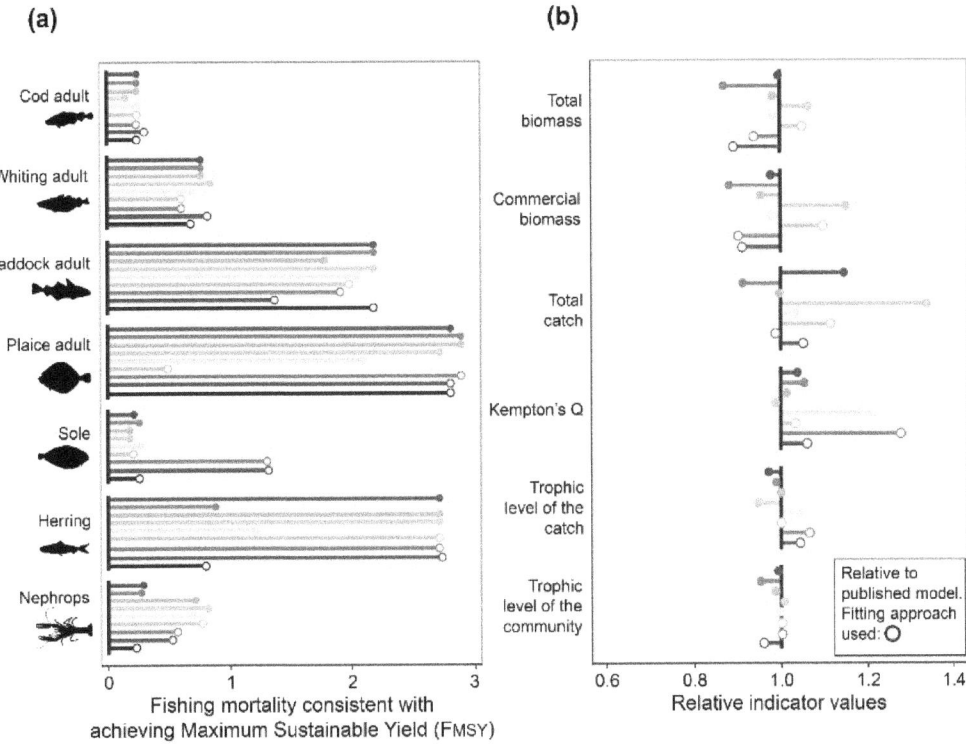

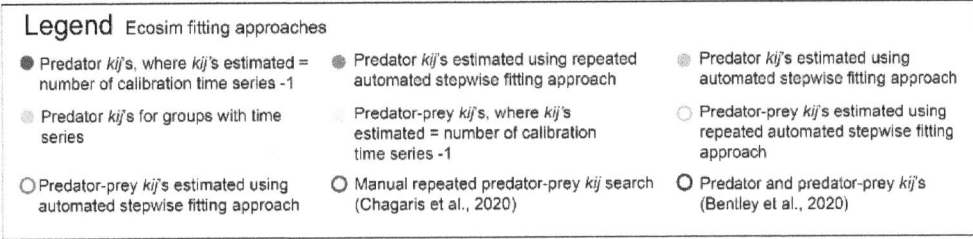

Figure 1. Irish Sea EwE outputs under alternate fitting approaches. Vulnerability multipliers for the Irish Sea Ecosim model were estimated following seven alternate fitting approaches. The impacts of alternate fitting approaches and vulnerability multiplier estimates are shown for (a) estimates of F_{MSY} (b) indicators of ecosystem structure and function. The impacts of vulnerability multiplier estimates on indicator simulations are illustrated by comparing new simulations against the simulations from the published model. The published Irish Sea model has vulnerability multiplier values estimated using the predator and predator-prey approach.[6]

(a)

(b)

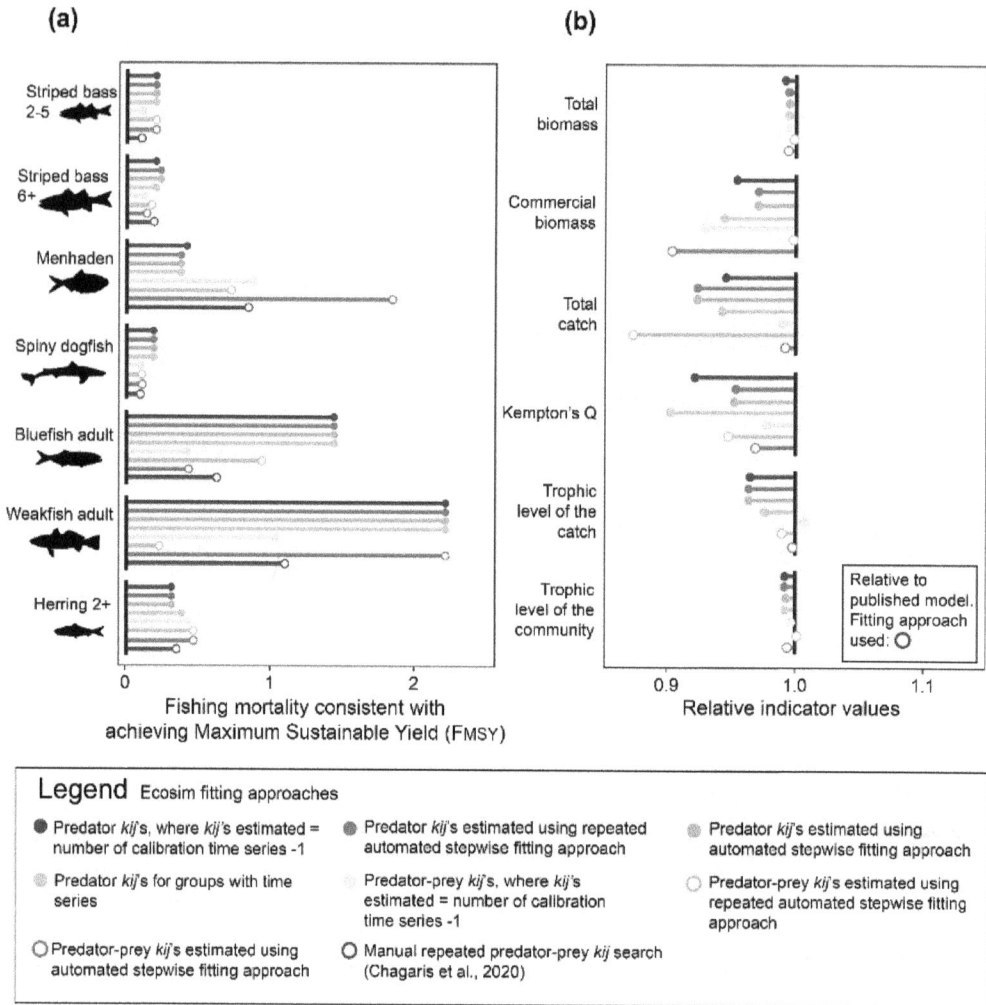

Figure 2. NWACS-MICE EwE outputs under alternate fitting approaches. Vulnerability multipliers for the NWACS-MICE Ecosim model were estimated following seven alternate fitting approaches. The impacts of alternate fitting approaches and vulnerability multiplier estimates are shown for (a) estimates of F_{MSY} (b) indicators of ecosystem structure and function. The impacts of vulnerability multiplier estimates on indicator simulations are illustrated by comparing new simulations against the simulations from the published model. The published NWACS-MICE model has vulnerability multiplier values estimated using the manual repeated predator-prey vulnerability multiplier search approach.[7]

Attribution This chapter is based on Bentley JW, Chagaris D, Coll M, Heymans JJ, Serpetti N, Walters CJ and Christensen V. 2024. Calibrating ecosystem models to

Media Attributions

- From Bentley et al. 2024. Table 1
- From Bentley et al. 2024. Figure 7
- From Bentley et al. 2024. Figure 8

Notes

1. Bentley, J.W., Serpetti, N., Fox, C.J., Heymans, J.J. and Reid, D.G., 2020. Retrospective analysis of the influence of environmental drivers on commercial stocks and fishing opportunities in the Irish Sea. Fisheries Oceanography, 29(5), pp.415-435. https://doi.org/10.1111/fog.12486

2. Chagaris, D., Drew, K., Schueller, A., Cieri, M., Brito, J. and Buchheister, A., 2020. Ecological reference points for Atlantic menhaden established using an ecosystem model of intermediate complexity. Frontiers in Marine Science, 7, p.606417. https://doi.org/10.3389/fmars.2020.606417

3. Howell, D., Schueller, A.M., Bentley, J.W., Buchheister, A., Chagaris, D., Cieri, M., Drew, K., Lundy, M.G., Pedreschi, D., Reid, D.G. and Townsend, H., 2021. Combining ecosystem and single-species modelling to provide ecosystem-based fisheries management advice within current management systems. Frontiers in Marine Science, 7, p.607831. https://doi.org/10.3389/fmars.2020.607831

4. Chagaris et al. 2020. *op. cit.*

5. See Supplemental figure 3 in Bentley et al. 2024

6. Bentley et al., 2020. *op. cit.*

7. Chagaris et al., 2020. *op cit.*

34.

Vulnerability multipliers: Discussion

Jacob Bentley; David Chagaris; Marta Coll; Sheila JJ Heymans; Natalia Serpetti; Carl J. Walters; and Villy Christensen

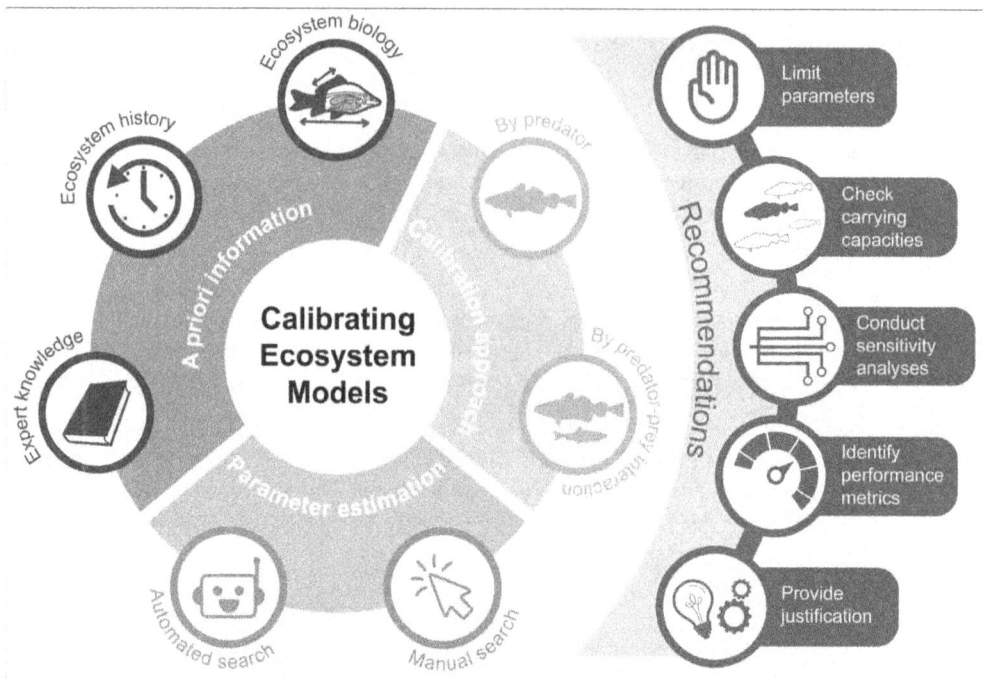

Limitations and future development

It is not enough to estimate vulnerability multipliers and assume those which produce the best statistical hindcast fit are appropriate. Ecological reasoning and hypothesis testing must support statistical inference as it should when balancing Ecopath models[1](Link 2015), estimating primary production anomalies[2] and incorporating environmental drivers.[3] Part of this process should include a critical evaluation of calibration time series, as these, and their associated uncertainty, drive the statistical estimation of vulnerability multipliers. Using data with inherent inconsistencies will lead to variable and potentially biased estimates. Equally important is the lack of missing reference time series. Time series produce constraints, and when estimating vulnerability multipliers for groups without time series the lack of constraints allows the fitting procedure to explore a broad parameter space to let such groups indirectly impact other groups with time series. All of the above, can impact model derived management advice.

As shown in the case studies in the previous chapters (#1, #2), F_{MSY} estimates can change in response to estimated vulnerability multipliers and their impacts on predator consumption rates, albeit with most changes being relatively conservative[4]. With high vulnerability multiplier values, species are more sensitive to changes in F and are therefore capable of recovering faster in the absence of fishing pressure, meaning maximum sustainable yields are achieved at lower fishing pressures (Figure 1a). As predator consumption rates increase with higher vulnerability multiplier values, prey experience higher predation rates, reducing the yield that can be obtained by fishing at F_{MSY} (Figure 1b). Unreliable vulnerability multipliers are not easily apparent when reviewing model hindcast simulations against calibration time series data. Comparing Ecosim F_{MSY} to other estimates, or proxies (e.g., natural mortality), is one approach to assess vulnerability multipliers and has been demonstrated for the ICES key-run models of the North Sea[5] and Irish Sea[6]. Simulating models beyond observations, under alternate fishing or environmental scenarios, can also highlight issues with vulnerability multipliers by exploring group sensitivities and whether simulated responses to change falls outside of what might be considered ecologically reasonable. Future developments should also consider dependencies between vulnerability multipliers, whether correlation exists between vulnerability multipliers, and how this may impact the ability of a search routine to find stable solutions.

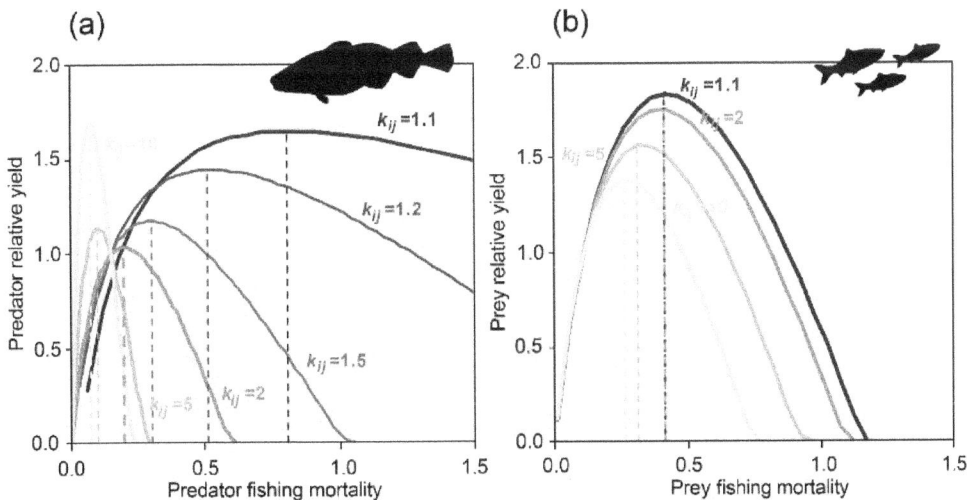

Figure 1. Effects of vulnerability multipliers on derived sustainable fishing advice. Estimations of fishing mortality at which MSY is achieved (F_{MSY}). Based on Walters et al., 2005.[7]

For EwE models to be of operational use, it should be possible to explain why estimated vulnerability multipliers are realistic. This could be based on knowledge of species' ecology, carrying capacity, or natural mortality. We envisage that the future development of Ecosim will encourage users to think more critically when calibrating models by building options to restrict the statistical vulnerability optimisation routine. The objective of this would be to enable users to constrain vulnerability multiplier estimation using *a priori* knowledge where, importantly, data is available to justify doing so. Increased control over the search for vulnerability multipliers could be used to set upper and lower parameter limits, or limits determined by carrying capacity, and penalise parameter combinations which operate outside of predefined limits.

Such constraints would also have important implications for Ecospace: the spatial-temporal component of EwE. High vulnerability multipliers, and the large increases in predation mortality which they enable, can have disproportionately large impacts in Ecospace when prey are restricted to small areas (as predators are able to deplete them rapidly). Vulnerability multipliers in Ecospace require further consideration given how spatial heterogeneity may impact species physiology, habitat carrying capacity, and predator-prey interaction rates. Spatial considerations are implicit within the vulnerability concept and enable spatial considerations to be integrated indirectly into Ecosim. The necessity for vulnerability multipliers, or at least those in Ecosim which go some way to indirectly recognising spatial heterogeneity, may be negated by the explicit consideration of spatial heterogeneity in Ecospace. Alternate vulnerability multiplier combinations may be needed depending on the priority use of Ecosim or Ecospace and the different mechanistic role vulnerability multipliers may play across the two components.

Recommendations

Calibration methods for EwE are not prescriptive. Any one of the methods included in Figure 2 of the Vulnerability and vulnerability multipliers chapter, or new methods, may be suitable for use if they can be justified. That said, the first case study showed that predator vulnerability multipliers are more likely to re-emerge than predator-prey vulnerability multipliers, and that re-emergence is impacted by data quality. Below we provide best practice recommendations to evaluate the appropriateness of vulnerability multipliers and their impact on model uncertainty:

- **Recommendation 1: Limit the number of vulnerability multipliers to be estimated.** The most efficient way to limit the number of parameters is to estimate by predator, add individual predator-prey combinations if you have specific arguments for why this is necessary. Avoid estimating vulnerabilities for groups without time series as the lack of constraints can lead to unrealistic estimates.

- **Recommendation 2: Explain vulnerability multipliers.** Provide justifications for setting initial vulnerability multipliers (or keeping the default). If estimating vulnerability multipliers using a statistical routine, check if they make sense relative to the exploitation and ecology of the predator and the ecology of the predator-prey interaction.

- **Recommendation 3: Sense check carrying capacities.** Vulnerability multipliers augment the upper limit for predator consumption rates, which dictates how predators respond to changes in mortality rates (e.g., release from fishing pressure or predation) or in prey biomass. It is important to review how predator biomass responds to such changes and critically evaluate whether the changes are plausible and whether the limits of estimates should be constrained (i.e., setting upper and lower limits).

- **Recommendation 4: Look beyond goodness of fit when evaluating model performance.** Combinations of vulnerability multipliers that achieve the best statistical fit (i.e., SS and AIC_c) do not necessarily produce the "best" model, if other model outputs, such as indicators, F_{MSY} reference points, and forward projections, are unlikely. Assessment of wider model performance should be undertaken to review vulnerability multipliers.

- **Recommendation 5: Perform vulnerability multiplier sensitivity analyses.** It is best practice to acknowledge and communicate model uncertainty. Calibrating Ecosim models, and thereby choosing one of multiple approaches to estimate vulnerability multipliers, introduces additional uncertainty into the process. Exploring model performance under alternate calibration approaches tests the sensitivity of model outputs to changes in vulnerability multipliers and identifies which vulnerability multipliers consistently emerge.

"

Attribution This chapter is based on Bentley JW, Chagaris D, Coll M, Heymans JJ, Serpetti N, Walters CJ and Christensen V. 2024. Calibrating ecosystem models to support marine Ecosystem-based Management. ICES Journal of Marine Science, https://doi.org/10.1093/icesjms/fsad213. Adapted based on CC BY License. Rather than citing this chapter, please cite the source.

Media Attributions

- Calibrating Ecosystem Models figure by Jacob Bentley
- From Bentley et al. 2024. Figure 9

Notes

1. Link, J.S. 2010. Adding Rigor to Ecological Network Models by Evaluating a Set of Pre-balance Diagnostics: A Plea for PREBAL. Ecol. Model. 221:1582-1593. 10.1016/j.ecolmodel.2010.03.012,

2. e.g., Serpetti N, Baudron AR, Burrows M, Payne BL, Helaouët P, Fernandes PG, Heymans J (2017) Impact of ocean warming on sustainable fisheries management informs the Ecosys-

tem Approach to Fisheries. Scientific Reports 7:13438 https://doi.org/10.1038/s41598-017-13220-7

3. e.g., Mackinson S. 2014. Combined analyses reveal environmentally driven changes in the North Sea ecosystem and raise questions regarding what makes an ecosystem model's performance credible? CJFAS. https://doi.org/10.1139/cjfas-2013-0173)

4. See Supplemental Figure 3 in Bentley et al. 2024

5. ICES. 2016. Report of the Working Group on Multispecies Assessment Methods (WGSAM), 9–13 November 2015, Woods Hole, USA. ICES CM 2015/SSGEPI:20. 206pp.

6. ICES. 2019b. Working group on multispecies assessment methods (WGSAM). ICES Scientific Reports. 1:320. Doi: 10.17895/ices.pub.5758

7. Walters CJ, Christensen V, Martell SJ, Kitchell JF. 2005. Possible ecosystem impacts of applying MSY policies from single-species assessment. ICES Journal of Marine Science 62:558 - 568. https://doi.org/10.1016/j.icesjms.2004.12.005

35.

Stock reduction analysis

A very useful technique for using long term data in stock assessment is Kimura's "stock reduction analysis". In this technique, historical catches are treated as fixed, known quantities, and are subtracted from simulated stock size over time so as to aid in estimating how large (and/or productive) the stock must have been in order to have sustained those catches and to have been reduced by some estimated fraction from its historical level. In some assessment literature, treating catches as fixed knowns is also called "conditioning on catch". A drawback of treating catches as fixed values is that catches in fact arise from the interaction of fishing effort and abundance, and ignoring this dynamic interaction amounts to treating the catches as purely depensatory impacts on stock size (when simulated stock size declines, the fixed catches can cause progressively larger calculated fishing mortality rates F, leading to a depensatory spiral of rapid collapse in the simulated stock, which may or may not have been possible in the real system).

When creating historical reference CSV files for model testing (see Import time series), all or part of a catch time series for any group(s) can be treated as a forcing input (with simulated F calculated each year as (input catch)/(simulated stock size)) by setting its data type to -6 (rather than the usual 6 for fitting catch data). Note that the catch time series for a group can be entered in two columns, with one column set to data type 6 and one to data type -6, where catches for years to be treated as forcing are placed in the -6 column and catches for years when catch is to be predicted from effort or assessment Fs placed in the 6 column. Most often, this splitting of catches into two columns should be used in cases where there are no independent assessments of F for some early years.

The Monte Carlo simulation interface in Ecosim can be used to search for Ecopath biomasses needed to have sustained historical catches. We cannot search for such initial biomass values by simple nonlinear search methods, due to the biomass constraints implied by Ecopath mass balance. The Monte carlo simulation interface can do a large number of simulations with randomly varying trial values of Ecopath biomasses, and can retain trial values that result in improved model fit; such a search or fitting procedure is known as a "Matyas search".

36.

Constrained optimization of fishing effort

Optimization methods like Ecosim's policy optimization search procedure (see Fishing policy exploration chapter) have been used to find, by fleet, fishing efforts that maximize various multi criteria benefit functions for ecosystem management, with criteria ranging from total profits to total employment and maintenance of ecosystem structure. These optimization methods work by running an ecosystem model that includes fishing mortality rates for multiple biomass groups by the multiple fleets, for long enough simulation periods for production and trophic interaction effects to play out. Fishing efforts by fleet are varied across simulation runs so as to seek the effort combination that maximizes the long-term benefit function. The result is a vector $EOPT_j$, $j=1,...,nf$, of optimum long-term efforts for the nf fleets included in the model.

Unfortunately, these optimum long-term efforts typically cannot be used directly in management strategy evaluations (MSEs), since they ignore constraints associated with fishery development (how fast efforts can grow or be reduced) and more importantly typically involve fishing efforts that would cause ecosystem simplification (overfishing of weaker stocks, even use of fleets to cull some species so as to increase productivity of others). In at least some jurisdictions like the United States, there is a strict legal mandate prohibiting over-harvesting (fishing rate exceeding F_{msy}) for any species. Further, in modeling management decision making over time, it is often necessary to represent the use of "control rules" that prescribe reduced fishing mortality rates for particular stocks when those stocks are below desired reference biomass levels.

A relatively simple linear programming method suggested by Murawksi and Finn[1] can be used to find short-term (annual) efforts that are constrained to be at or below the long-term optima $EOPT_j$, while respecting the relative fishery values represented in the long-term optimization and also constraints associated with fishing mortality rate targets or upper limits by species. Suppose that the fleets harvest $i=1,...,ns$ species (or biomass groups), with each unit of effort for fleet j causing a fishing mortality rate q_{ij} on species i (q_{ij} is the catchability coefficient for species i by fleet j, and is also the Ecopath base fishing mortality rate for i,j). Suppose that for each species i, there is a target or maximum allowable fishing mortality rate $FTARGET_i$, summed over all fleets that cause fishing mortality (landings and/or discards) on i. Suppose that we assign an "importance weight" v_j to effort by fleet j, where v_j reflects the relative value of increasing (or maintaining) effort for fleet j because of its contribution to overall long term ecosystem value and/or its legal entitlement to fish. Then the linear programming optimization can be formulated simply, as, maximize,

$$\sum_{j=1}^{nf} v_j E_j \tag{1}$$

by varying the E_j subject to the constraints,

$$EMIN_j < E_j < EOPT_j \text{ for } j=1, ..., nf \tag{2}$$

$$\sum_{j=1}^{nf} q_{ij} E_j \leq FTARGET_i \text{ for } j=1, ..., nf \tag{3}$$

That is, try to make the efforts E_j as large as possible without exceeding $EOPT_j$, while allowing efforts of at least $EMIN_j$ and not allowing the sums of $q_{ij}E_j$ to exceed target fishing rate for any species i. An alternative formulation for recognizing fishery development rate constraints would be to replace the first set of nf constraints (Eq. 2) with $EMIN_j < E_j < EMAX_j$. Where the $EMIN$ and $EMAX$ are allowed to vary from year to year by limited increments from the previous year's values, and are not allowed to exceed $EOPT$.

Figure 1 illustrates the kind of complicated solutions that can arise from this optimization, even for a very simple case where two fishing fleets pursue just two species, with one fleet having higher catchability for one of the species and the other fleet having higher catchability for the other species.

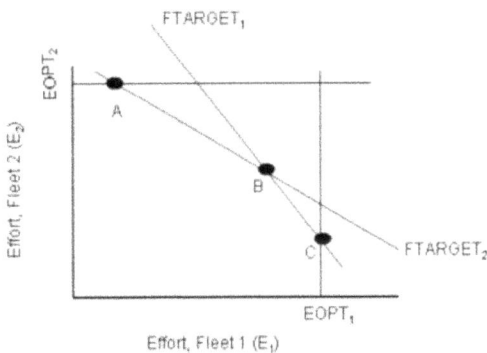

Effort, Fleet 1 (E_1)

Figure 1. Graphical representation of the linear programming problem. Each line represents a constraint (vertical and horizontal lines are the *EOPT* constraints, sloped lines are the *FTARGET* constraints.

The lines in Figure 1 represent effort levels that exactly meet the constraints; efforts must be to the left and below each line in order to be feasible. Thus the feasible effort combinations are only those in the polygon from the graph origin out to the first constraint lines met. Since efforts are to be as large as possible, the solution has to lie along one of those first lines met, and in fact has to be on one of the three vertices marked A, B, C.

- Effort combination A represents Fleet 1 being severely restricted, but Fleet 2 operating at its *EOPT.*

- Effort combination C represents Fleet 2 being severely restricted but Fleet 1 operating at its *EOPT,* and

- Effort combination B represents a "balanced" policy choice where both fleets are restricted to below *EOPT* in order to "share the burden" of avoiding exceeding either of the two *FTARGET* species constraints.

Which of these three combinations will be chosen (solve the linear programming maximization) depends on the value weights v_j in Eq. 1. If v_1 is much higher than v_2, combination C will be chosen, combination A will be chosen if v_2 is larger, and setting the two v's equal will more likely lead to the balanced combination B. Note also in Figure 1 that as the *FTARGET* constraints are "relaxed" (increased so the sloped lines move upward and to the right), it becomes more likely that the optimum effort combination will lie near *EOPT* for both fleets; likewise, as these constraints are "tightened" (reduced so the sloped lines move down and to the left), it becomes more likely that the *EOPTs* will not be in the feasible region so that the optimum solution will lie either with a mixed effort combination or with one or another of the fleets shut down entirely.

Using the linear programming formulation, it is simple to evaluate the cost, in terms of lost total value, of introducing more restrictive constraints on species harvest rates. In the overall management strategy evaluation setup for Ecosim, the only other way to evaluate this cost is to do policy runs with and without "weakest stock" constraints on fleet quotas, where all fleets are assumed to share equally in reductions needed to meet such constraints. The linear programming solution may well demonstrate that such equal sharing of the conservation burden is in fact far from optimum.

The linear programming formulation can also be used to demonstrate potential increases in fishery value from selective fishing practices that change the species-specific catchabilities qij. For example, if q_{12} and q_{21} (catchability of species 1 by fleet 2 and of species 2 by fleet 1) could be greatly reduced in the Figure 1 example, the slopes of the two *FTARGET* constraint lines would decrease/increase so as to move the solution toward higher total efforts (move point B up and to the right, closer to the *EOPT1-EOPT2* intersection) and thus higher total value.

The key to getting useful results from the linear programming exercise is to make wise choice of the fleet value weights v_j. One objective option for doing this each year (assuming a management strategy where biomass and perhaps catchability q_{ij} estimates are being updated regularly) is to set each weight to be

$$v_j = \sum_{i=1}^{nf} P_{ij} q_{ij} B_i \qquad (4)$$

where P_{ij} is the landed price for species i by fleet j and B_i is the current estimated biomass of species i. Using this formula, v_j is just the sum over species of predicted catches per effort times prices, so that $v_j B_j$ represents the (short term) predicted total value of landings by fleet j and the overall linear programming objective function just becomes the predicted total landed value of all catches. This option can lead to complex policy changes over time when the *FTARGET*i decrease with decreases in biomasses B_i. A simpler version of Eq. 4 would be to use Ecopath base or estimated unfished biomasses and catchabilities rather than current estimates, so as to have the weights not change over time.

Notes

1. Murawski, S.A. and J.T. Finn. 1986. Optimal effort allocation among competing mixed-species fisheries, subject to fishing mortality constraints. Can. J. Fish. Aquat. Sci. 43: 90-100. https://doi.org/10.1139/f86-010

37.

Management strategy evaluation

Management strategy evaluation (MSE) is concerned with evaluating uncertainty about the impact of applying alternative formal rules for varying fisheries over time, given uncertainty about ecosystem dynamics (as represented by key Ecopath and Ecosim parameters) and about biomass states over time due to stock assessment errors. The main tool in EwE for simulating such uncertainties and rules is the "CEFAS MSE plug-in" *(Tools > Cefas MSE)* developed by Mackinson and colleagues.[1]

The CEFAS plugin provides a way to represent uncertainty by doing multiple simulation scenarios with different parameter values, while comparing various management performance measures across these simulations for alternative harvest management decision rules. The decision rules included in the plug-in were developed in response to questions raised about multi-species management strategies in the ICES (North Sea) management area, but broadly represent harvest management decision rule options for most highly managed areas in the world.

Management decision rules are represented by several main components:

1. So-called "harvest control rules" (HCRs, see Figure 1) representing desired patterns of variation in total fishing mortality rate for target species as a function of species biomass (comparisons of such HCRs is a central concern in single-species management strategy evaluation).

2. Imposition of alternative rules for discarding, e.g. forbidding discarding may effectively cause "weakest stock" management by not allowing fishing fleets to avoid harvest limits by discarding catches of weaker stocks), or allowing highest value management by seeking fleet operations that will maximize total profit despite discarding impacts, or assuming ability of fleets to do "selective fishing" of only valued target species.

3. Limits on total fishing effort by fleets and on year-to-year allowable changes in fishing effort.

Quite complex calculations are done at each simulation time step in order to meet (or not) the constraints implied by the single-species HCRs. For details, see the supporting information from Mackinson et al.[2]

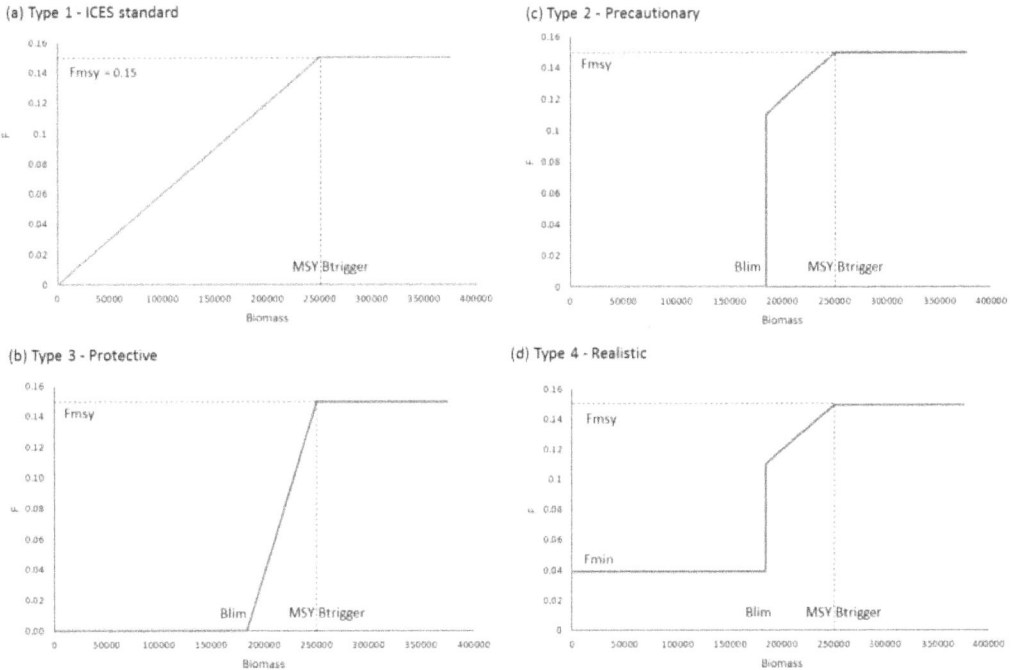

Figure 1. Examples of harvest control rules (HCRs) used for management strategy evaluation. Type 1 is the "ICES standard" advice rule where F declines linearly to zero when biomass is below *MSY B*_{trigger}. Type 2 is "Precautionary", easing the rate of reduction in *F* between *MSY B*_{trigger} and *B*_{lim}. Type 3 is the most "Protective", applying the most severe reductions in *F*, which declines linearly from *F*_{msy} at *MSY B*_{trigger} to zero at *B*_{lim} . Type 4 is considered the most "Realistic", similar to Precautionary, but recognizes that a small level of residual non-target by-catch mortality may remain on a stock at *B* ≤ *B*_{lim}. From Mackinson et al. (*op. cit.*, Figure 4).

A key part of management strategy evaluation is to determine the effect on management performance of errors in biomass estimates due to random survey variation and to cumulative errors caused by stock assessment procedures. For multispecies models, it is not practical to simulate the complex data gathering and assessment model fitting procedures that are used to obtain biomass estimates for more valuable species. Instead, the plug-in uses an observation by Walters[3] that errors in biomass estimates from common stock assessment methods (virtual population analysis, statistical catch at age analysis or stock synthesis models) propagate over time in a relatively simple pattern that can be modelled by a statistical filtering equation. The equation predicts that even single large survey errors can cause errors in stock assessments (impacts on estimates from assessments) that propagate over several years so as to possibly have large cumulative impacts on over- and/or under-harvesting.

Figure 2. Flowchart indicating how the CEFAS MSE plug-in first creates a (large) number of possible EwE models (drawn randomly) then continues to the management strategy evaluation (Figure 3). From Mackinson et al. (*op. cit.*, Figure 1)

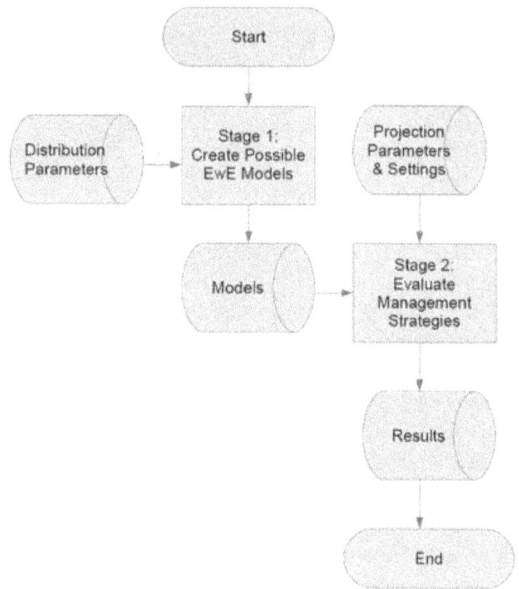

The CEFAS MSE plug-in approach has two stages. First it creates a number of plausible EwE (Ecopath) models by sampling distributions of input parameters (biomass, feeding, production and consumption rates and predator-prey interaction rates) (Figure 2). Next, it simulates the effect of alternative management strategies defined by their HCR and regulatory mechanisms (Figure 3).

For guidance of how to use the CEFAS MSE plug-in, most notably see the paper and supplementary materials for Mackinson et al. (*op. cit.*) as well as the tutorial on this (web and pdf versions only).

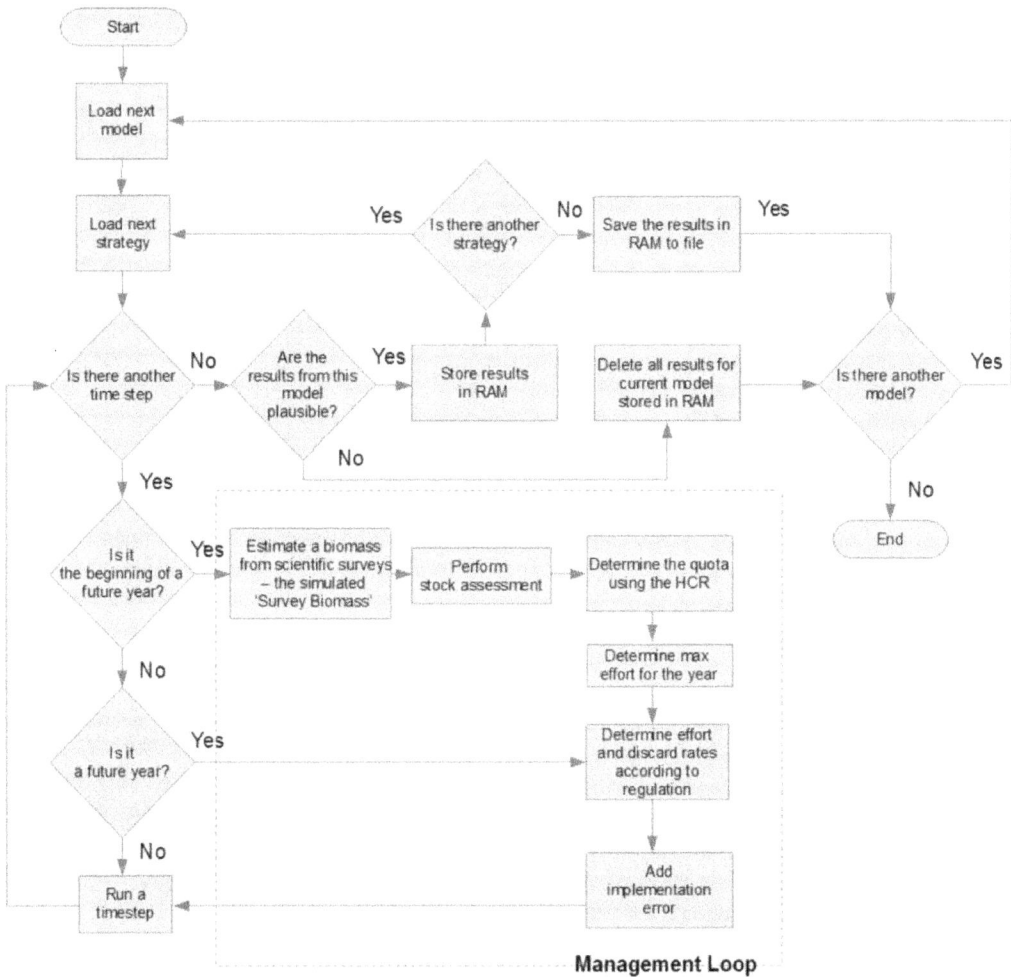

Figure 2. Flowchart for evaluation of alternative management strategies. From Mackinson et al. (*op. cit.*, Figure 2).

Notes

1. Mackinson S, Platts M, Garcia C, Lynam C (2018) Evaluating the fishery and ecological consequences of the proposed North Sea multi- annual plan. PLoS ONE 13(1): e0190015. https:// doi.org/10.1371/journal.pone.0190015

2. Mackinson et al. (2018) S1 Supporting Information. Technical methods of the uncertainty and MSE routine. https://doi.org/10.1371/journal.pone.0190015.s001

3. Walters C. 2004. Simple representation of the dynamics of biomass error propagation for stock assessment models. CJFAS 61(7):1061-1065. https://doi.org/10.1139/f04-120

38.

Fleet effort dynamics

Ecosim and Ecospace can include fishing pressure in two ways: using fishing mortality or fishing effort. If fishing mortality is used, the corresponding catch is calculated for each time step from catch = fishing mortality · biomass. If effort is used, the key assumption is that the fishing mortality in the Ecopath baseline model corresponds to an effort of 1 (unity). Any change in effort over time will result in a proportional change in fishing mortality.

So, if in the Ecopath baseline, catch = 0.2 t km^{-2} year^{-1} and biomass for the group in question is 1 t km^{-2}, the fishing mortality is calculated from the catch/biomass ratio to 0.2 year^{-1}. If fishing effort increases, e.g., to 1.1 then this results in an F of 0.2 · 1.1 = 0.22 year^{-1}. This is not an EwE invention, it follows straight from how fishing effort was originally defined.[1]

Effort is associated with fleets, and many fleets catch more than one species. That's fine, the F's will show the same proportional change for all species. But what if both effort for a fleet impacts a species for which there also a fishing mortality entered? In that case we have no choice but to let the fishing mortality overrule the effort for such a species. This indeed offers some flexibility, for instance in an application with a multi species fleet where there's detailed information from assessment for one species. We can then use the fishing mortality from the assessment for that species, and fleet effort for the rest.

There are three ways to specify temporal changes in fishing fleet sizes and fishing effort:

1. By sketching temporal patterns of effort in the model run interface;

2. By entering annual patterns via reference CSV files along with historical ecological response data; and

3. By treating dynamics of fleet sizes and resulting fishing effort as unregulated and subject to fisher investment and operating decisions ("bionomic" dynamics, fishers as dynamic predators).

To facilitate exploration of alternative harvest regulation policies, the Ecosim default options are (1) or (2). However, you can invoke the fleet/effort dynamics model where effort is estimated, rather than input, by checking *Ecosim > Input > Ecosim parameters > Fleet effort dynamics*. Input parameters must then be set on the *Ecosim > Input > Fleet size dynamics* form.

When the fleet/effort response option is invoked, Ecosim starts each by erasing all previously entered time patterns for fishing efforts and fishing rates, and replaces these with simulated values generated as each simulation proceeds. The fleet/effort dynamics simulation model uses the idea that there are two time scales of fisher response:

1. A short time response of fishing effort to potential income from fishing, within the constraints imposed by current fleet size, and

2. A longer time investment/depreciation "population dynamics" for capital capacity to fish (fleet size, vessel characteristics).

These response scales are represented in Ecosim by two "state variables" for each gear type *g*.

Fast time response model

$E_{g,t}$ is the current amount of effort (active, searching gear, scaled to 1.0 at the Ecopath base fishing mortality rates), and $K_{g,t}$ is the maximum possible effort ($E_{g,t} < K_{g,t}$).

At each time step, a mean income per effort index $I_{g,t}$ is calculated as

$$I_{g,t} = \sum_{i} q_{g,i} B_i P_{g,i} \tag{1}$$

where i = ecological species or biomass group, $q_{g,i}$ is the catchability coefficient (possibly dependent on B_i) for species i by gear g, and $P_{g,i}$ is the market price obtained per biomass of i by gear g fishers. Also, mean fleet profit rates $PR_{g,t}$ for fishing are calculated thus:

$$PR_{g,t} = I_{g,t} - c_g \tag{2}$$

where c_g is the cost of a unit of fishing effort for gear g (cost and price factors are entered via the Definition of fleets and Market price forms). For each time step, the "fast" effort response for the next (monthly) time step is predicted by a sigmoid function of income per effort and current fleet capacity:

$$E_{g,t+1} = \frac{K_{g,t} I_{g,t}^p}{I_{hg}^p + I_{g,t}^p} \tag{3}$$

Here, I_{hg} and p are fleet-specific response parameters. I_{hg} is the income level needed for half maximum effort to be deployed and p is a "heterogeneity" parameter for fishers: high p values imply all fishers "see" income opportunity similarly (start or quit at similar income values), while low p values imply fishers "turn on" their effort over a wide range of mean incomes (start or quit over a wide range of average incomes), as shown in Figure 1.

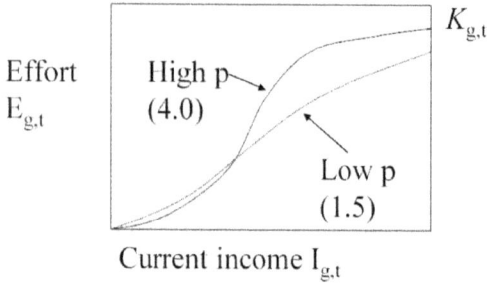

Figure 1. Effect of the "heterogeneity" parameter, *p*, on effort/income function.

Slow time response model

For each fleet, slow effort responses are modelled as changes in fleet capacity ($K_{g,t}$), which is a function of the capital depreciation rate ρ_g, the capital growth rate $r_{g,t}$ and profit $PR_{g,t}$. The capital growth rate is calculated via a growth factor $gf_{g,t}$, i.e.,

$$gf_{g,t+1} = \frac{K_{g,t}(r_{g,t} + \rho_g)}{PR_{g,t}} \quad (4)$$

where $K_{g,1}$, ρ_g and $r_{g,1}$ are set by the user. The annual capacity $K_{g,t}$ is then updated as

$$K_{g,t+1} = K_{g,t}(1 - \rho_g) + gf_{g,t}PR_{g,t+1} \quad (5)$$

Attribution This section was adapted and edited from the unpublished 2008 EwE User Guide.

Notes

1. Beverton, R.J.H. and Holt, S.J. 1957. On the dynamics of exploited fish populations. Fisheries Investigations, 19, 1-533. Chapman and Hall, Facsimile reprint 1993, London. 533 pp.

VI

Ecological, Social and Economic Factors

39.

Introduction

Santiago de la Puente

> If you need a refresher in economics, see, e.g., the open source textbooks Principles of Macroeconomics and Principles of Microeconomics.

Fisheries systems are composed of linked biophysical and human subsystem with interacting ecological, economic, social, and cultural components.[1] In previous and subsequent sections of this book, information is provided on how to use EwE models to assess the ecological consequences of environmental change as well as of fisheries policies, conservation efforts (e.g., marine protected areas), and alternative uses of the marine environment (e.g., off-shore wind farms). Yet, humans, being integral parts of fisheries systems, also require attention in our modelling efforts.

Fishers' actions are driven by economic objectives and their desire to secure their well-being.[2][3] Thus, their choices and behaviours, whether legal or not, occur along a gradient and are implemented by the same individuals or vessels, in their pursuit for a good life.[4] Understanding the socio-economic characteristics of the fishers and fleets operating in the systems we are modelling, is key to assess whether the economic incentives at play promote compliance with existing or proposed regulations.[5][6][7][8] Moreover, environmental and policy changes have direct and indirect consequences on fishers' income and broader social and economic policy goals. Yet, managers, decision makers, and regulators often lack adequate performance metrics covering the human dimensions of fisheries systems. This limits their capacity to assess or predict the socio-economic consequences of change in fisheries systems, as well as trade-offs among conflicting management objectives and the uncertainty behind them.[9][10][11][12]

More efforts are needed to adequately characterize and couple fisheries and their linked downstream supply chain nodes in food web models so that they can be used to inform decisions in the realms of the ecosystem approach to fisheries, ecosystem-based fisheries management, ecosystem-based management.[13][14] [13,14]. Fortunately, EwE has multiple capabilities for including fisheries socio-economics in food web models and improving our understanding of the feedbacks between human activities and ecosystem dynamics.

Notes

1. Charles A., Human dimension in marine ecosystem-based management, in: M.J. Fogarty, J.J. McCarthy (Eds.), The Sea: Marine Ecosystem-Based Management, Harvard University Press, 2014: pp. 57–75. https://www.hup.harvard.edu/books/9780674072701

2. Sethi S.A., T.A. Branch, R. Watson, Global fishery development patterns are driven by profit but not trophic level, Proc. Natl. Acad. Sci. 107 (2010) 12163–12167. https://doi.org/10.1073/pnas.1003236107

3. Weeratunge N., C. Béné, R. Siriwardane, A. Charles, D. Johnson, E.H. Allison, P.K. Nayak, M.-C. Badjeck, Small-scale fisheries through the wellbeing lens, Fish and Fisheries 15 (2014) 255–279. https://doi.org/10.1111/faf.12016

4. Cisneros-Montemayor A.M., S. Harper, T.C. Tai, The market and shadow value of informal fish catch: a framework and application to Panama, Natural Resources Forum 42 (2018) 83–92. https://doi.org/10.1111/1477-8947.12143

5. Grafton R.Q., R. Arnason, T. Bjørndal, D. Campbell, H.F. Campbell, C.W. Clark, R. Connor, D.P. Dupont, R. Hannesson, R. Hilborn, J.E. Kirkley, T. Kompas, D.E. Lane, G.R. Munro, S. Pascoe, D. Squires, S.I. Steinshamn, B.R. Turris, Q. Weninger, Incentive-based approaches to sustainable fisheries, Canadian Journal of Fisheries and Aquatic Sciences 63 (2006) 699–710. https://doi.org/10.1139/f05-247

6. Grafton R.Q., T. Kompas, R. Hilborn, Economics of Overexploitation Revisited, Science 318 (2007) 1601–1601. https://doi.org/10.1126/science.1146017

7. Nøstbakken L., Fisheries law enforcement—A survey of the economic literature, Marine Policy 32 (2008) 293–300. https://doi.org/10.1016/j.marpol.2007.06.002

8. Diekert F., L. Nøstbakken, A. Richter, Control activities and compliance behavior—Survey evidence from Norway, Marine Policy 125 (2021) 104381. https://doi.org/10.1016/j.marpol.2020.104381

9. Hilborn R,, Defining success in fisheries and conflicts in objectives, Marine Policy 31 (2007) 153–158. https://doi.org/10.1016/j.marpol.2006.05.014

10. Stephenson R.L., A.J. Benson, K. Brooks, A. Charles, P. Degnbol, C.M. Dichmont, M. Kraan, S. Pascoe, S.D. Paul, A. Rindorf, M. Wiber, H. editor: L. Pendleton, Practical steps toward integrating economic, social and institutional elements in fisheries policy and management, ICES Journal of Marine Science 74 (2017) 1981–1989. https://doi.org/10.1093/icesjms/fsx057

11. Hilborn R., E.A. Fulton, B.S. Green, K. Hartmann, S.R. Tracey, R.A. Watson, When is a fishery sustainable?, Canadian Journal of Fisheries and Aquatic Sciences 72 (2015) 1433–1441. https://doi.org/10.1139/cjfas-2015-0062

12. Punt A.E., Strategic management decision-making in a complex world: quantifying, understanding, and using trade-offs, ICES Journal of Marine Science 74 (2017) 499–510. https://doi.org/10.1093/icesjms/fsv193

13. Nielsen J.R. et al., Integrated ecological-economic fisheries models-Evaluation, review and challenges for implementation, Fish and Fisheries 19 (2018) 1–29. https://doi.org/10.1111/faf.12232

14. Craig J.K., J.S. Link, It is past time to use ecosystem models tactically to support ecosystem-based fisheries management: Case studies using Ecopath with Ecosim in an operational management context, Fish Fish (2023). https://doi.org/10.1111/faf.12733

40.

Revenue and profits

Santiago de la Puente

EwE can directly produce multiple socio-economic indicators. Nonetheless, EwE models can also be linked to, or coupled with, external bioeconomic models to expand its capabilities. A first step, however, is the identification of ex-vessel prices for the multiple functional groups included in the model. These data can be obtained from local or national reports, as well as from existing regional (*e.g.*, EU's Scientific, Technical and Economic Committee for Fisheries) and global (*e.g.*, Sea Around Us) databases.

Users should begin by defining the currency to be used in the model by selecting the correct monetary units (*e.g.*, USD, EUR, CNY) while setting the models' parameters (*Ecopath > Input > Model parameters*). Then, after defining the model's functional groups and fishing fleets, and including the corresponding landings data, users will be able to input price data for each functional group – fishing fleet combination with catches. This is achieved in the off-vessel prices form (*Ecopath > Input > Fishery > Off-vessel prices*), and the data included should reflect the value of a ton of fish or shellfish caught within the model area in the base year. Based on this data, the model can estimate three socio-economic indicators pertaining to revenue by multiplying the off-vessel prices with the corresponding landed amounts for each functional group – fishing fleet combination and adding these values across: (i) functional groups (i.e., the landed value per functional group), (ii) fishing fleets (i.e., fleet level revenue), and (iii) across functional groups and fishing fleets (i.e., total landed value or total producer revenue).

If multiple Ecopath models are available for the same project area, then the landed value per functional group, the fleet level revenue, and the total producer revenue can be used to compare the fisheries over time. Moreover, if time series data for off-vessel prices is available, then the outputs from Ecosim and Ecospace runs (e.g., time series data of catches per fleet) can be used to estimate these three indicators externally. Yet, EwE can also model these internally, if information is available on how sensitive off-vessel prices are to changes in landed quantities. This can be achieved through price elasticities (*Ecosim > Input > Price elasticity*). For more information on this see the chapter on Price elasticity.

In some cases, the value of a fishery may increase over time (due to higher catches or off-vessel prices). However, the fleets' profitability might not follow the same trend if fishing costs (e.g., wages, fuel) have increased at a much faster rate. Indicators related to profits (i.e., *Profits = Revenue – Costs*) are thus useful for capturing these changes, particularly if multiple fleets are involved. This is also a relevant concern for modelling the consequences policies pertaining to fisheries subsidies.[1]

There are multiple ways by which EwE allows users to get a grasp of profitability. It begins by obtaining information about the cost-income structures of the fishing fleets. These data can be directly collected using surveys or semi-structured interviews[2] [18], extracted from secondary literature (mainly grey literature), or requested to the government bodies responsible for its collection at subnational or national levels. EwE allows users to input cost-income data by expressing it as a percentage of the fleet level revenue, where: *Total Value of the fleet = Fixed costs + Cost per unit of effort + Cost of sailing + Profit*. This information can be included in the model in the same form used to define the fleets (*Ecopath > Input > Fishery > Fleets*).

Here, *Fixed costs* include costs that are independent of changes in effort levels at the fleet scale (*e.g.*, capital investments, management, and monitoring costs). *Cost per unit of effort* and *Cost of sailing* are both used to express variable costs, or costs that change proportionally to changes in fishing effort (e.g., fuel, food for the crew, wages). If the user only plans to use Ecopath or Ecosim, then all variable costs should be included as *Cost per unit of effort* (i.e., *Cost of sailing* should be left empty). Only if the intention is to build an Ecospace model, then variable costs should be split, highlighting the cost fraction that will vary directly depending on the spatial allocation of fishing effort (e.g., fuel costs). These costs should be included as *Cost of sailing*, while all other variable cost should be entered as *Cost per unit of effort*. The default settings for all fleets places *Fixed costs* at 0%, *Cost per unit of effort* at 40%, *Cost of sailing* at 40% and *Profit* at 20%. These defaults might be representative for some fleets, yet for most they will not (e.g., STECF 22-06).

Understanding what proportion of the fleet level revenues correspond profits allows users to estimate fleet level profits out of Ecosim and Ecospace runs.[3 4 5 6 7 8 9]. Moreover, if additional information is available of what percentage of the fleet level revenue is used for wages, how many vessels are operating and how many people are employed per vessel, then modelers can also estimate fishers' average salaries per fleet. This information could then be used to compare it with annual country level minimum wages to assess if, and under which scenarios, fishers operate below the poverty line. Additionally, having information about fishing costs and profits is essential for modelling effort dynamics within EwE (see the chapter on Fleet effort dynamics), as well as for fishing policy exploration (see chapter on Fishing policy exploration). An alternative way to parametrize fishing costs is described in the Value chain modelling chapter.

Notes

1. Sumaila U.R., N. Ebrahim, A. Schuhbauer, D. Skerritt, Y. Li, H.S. Kim, T.G. Mallory, V.W.L. Lam, D. Pauly, Updated estimates and analysis of global fisheries subsidies, Mar Policy 109 (2019) 103695. https://doi.org/10.1016/j.marpol.2019.103695

2. Bennett N.J., A. Schuhbauer, D. Skerritt, N. Ebrahim, Socio-economic monitoring and evalua-

tion in fisheries, Fish Res 239 (2021) 105934. https://doi.org/10.1016/j.fishres.2021.105934

3. Wang Y., S.Y. Li, L.J. Duan, Y. Liu, Fishery policy exploration in the Pearl River Estuary based on an Ecosim model, Ecol. Model. 230 (2012) 34–43. https://doi.org/10.1016/j.ecolmodel.2012.01.017

4. Ramírez A., M. Ortiz, J. Steenbeek, V. Christensen, Evaluation of the effects on rockfish and kelp artisanal fisheries of the proposed Mejillones Peninsula marine protected area (northern Chile, SE Pacific coast), Ecol. Model. 297 (2015) 141–153. https://doi.org/10.1016/j.ecolmodel.2014.11.012

5. Izquierdo-Gomez D., J.T. Bayle-Sempere, F. Arreguín-Sánchez, P. Sánchez-Jerez, Modeling population dynamics and small-scale fisheries yields of fish farming escapes in Mediterranean coastal areas, Ecol. Model. 331 (2016) 56–67. https://doi.org/10.1016/j.ecolmodel.2016.01.012

6. Bacalso R.T.M., M. Wolff, R.M. Rosales, N.B. Armada, Effort reallocation of illegal fishing operations: A profitable scenario for the municipal fisheries of Danajon Bank, Central Philippines, Ecol. Model. 331 (2016) 5–16. https://doi.org/10.1016/j.ecolmodel.2016.01.015

7. Rehren J., M. Wolff, N. Jiddawi, Holistic assessment of Chwaka Bay's multi-gear fishery – Using a trophic modeling approach, J. Mar. Syst. 180 (2018) 265–278. https://doi.org/10.1016/j.jmarsys.2018.01.002

8. Armada N.B., R.T.M. Bacalso, R.M.P. Rosales, A.T. Lazarte, Right-sizing as a strategy for allocating fishing effort in a defined marine ecosystem: A Philippines case study, Ocean Coast. Manag. 165 (2018) 167–184. https://doi.org/10.1016/j.ocecoaman.2018.08.018

9. Alms V., G. Romagnoni, M. Wolff, Exploration of fisheries management policies in the Gulf of Nicoya (Costa Rica) using ecosystem modelling, Ocean Coast. Manag. 230 (2022) 106349. https://doi.org/10.1016/j.ocecoaman.2022.106349

41.

Value chain modelling

Santiago de la Puente

Economist classify fisheries as primary industries, highlighting their supporting role for numerous secondary (*e.g.*, seafood processors) and tertiary industries (e.g., hotels and restaurants) in the economy.[1][2] Extractive industries are generally limited at creating value. For example, the direct economic contribution of fisheries to national economies around the world ranges only between 0.5% and 2.5% of their Gross Domestic Product (GDP).[3][4].

However, fishing requires inputs from other industries to operate (e.g., boat building industry and fishing net manufacturers),[5] and as fish, marine invertebrates and macroalgae move along supply chains, they are transformed into accessible seafood products tailored to meet the needs of consumers.[6] Thus, fisheries have "upstream" (i.e., prior to fishing) and "downstream"(i.e., post-harvesting) economic effects, which are commonly characterized using input-output models to estimate multipliers (see the External bio-economic models chapter). These multipliers are factors used for approximating the extent of the contribution of an economic sector to a nation's economy.

The EwE Value Chain plug-in is a powerful tool for characterizing the downstream economic effects of the fishing fleets.[7] The economic agents or components within the value chain are referred to as *enterprises* described via a set of common attributes (see Table 1) and segregated based on their function within the value chain. The value chain model is coupled to ecosystem model through the fishing fleets. These are considered as *producers*, given that they are the main source of raw materials for the seafood supply chain. Non-extractive activities using marine living resources (*e.g.,* ecotourism or non-retaining recreational fisheries) and some types of aquaculture enterprises can also be considered as *producers*. Enterprises that receive marine living resources from *producers* and transform them into seafood are classified as *processors*. These typically include fish cutters and filleters, canneries, seafood freezing facilities, or fishmeal processing plants.[8][9][10] However, some aquaculture operations can also be classified as *processors* if they receive feed or seeds from other *processors* or *producers* within the system.[11]

Table 1. Input parameters used to characterize seafood value chains in EwE's value chain plug-in.

Topic	Parameter	Symbol	Unit
Identity	Name		
	Nationality		
Products	Agricultural	R_a	$/t
	Energy	R_c	$/t
	Industry	R_i	$/t
	Services	R_s	$/t
	Ticket sales	R_t	$/effort
Subsidies	Energy	U_e	$/t
	Other	U_o	$/t
Pay or share	Worker, female	P_s or S_s	$/t or %
	Worker, male	P_h or S_h	$/t or %
	Owner, female	P_f or S_f	$/t or %
	Owner, male	P_m or S_m	$/t or %
Input	Agricultural	I_a	$/t
	Capital cost	I_c	$/t
	Energy cost	I_e	$/t
	Industrial cost	I_e	$/t
	Services cost	I_s	$/t
Cost	Management	C_m	$/t
	License	C_l	$/t
	Certification	C_c	$/t
	Observers	C_o	$/t
	Observer rate	C_r	
Taxes	Environmental	T_e	$/t
	Export	T_x	$/t
	Import	T_i	$/t
	Production	T_p	$/t
	Value added tax	T_v	$/t
	Licenses	T_l	$/t

Topic	Parameter	Symbol	Unit
	Worker, female	J_s	#/t
	Worker, male	J_h	#/t
Employment	Owner, female	J_f	#/t
	Owner, male	J_m	#/t
	Female worker dependents	D_s	#/worker
	Male worker dependents	D_h	#/worker
Dependents	Female owner dependents	D_f	#/owner
	Male owner dependents	D_m	#/owner

The remaining enterprises in the seafood value chain are classified as either *distributors* or *sellers*, depending on their functional roles. *Distributors* typically include middlemen linking *producers* and *processors*, enterprises specialized on transporting seafood from *processors* to *sellers*, or seafood exporters linking *producers* with foreign markets. *Sellers*, on the other hand, connect *producers*, *processors* and *distributors* with consumers, and these typically include seafood wholesalers, supermarkets, small municipal markets, street vendors and restaurants.

It is important to highlight that the information required to populate the value chain within EwE, such as an enterprise's management costs or its number of female workers (Table 1) must be expressed in a per tonne basis (e.g., $ per tonne or jobs per tonne). Moreover, the forms within the plug-in (*Ecopath > Output > Tools > Value chain*) provide ample freedom regarding what information to include. If available data for characterizing enterprises is highly aggregated, it is still possible to populate the value chain by including a single item in the cost structure. The same is the case for the employees, workers, and dependents if data is not segregated by sex.

Links between enterprises are characterized by tracking losses in weight and gains in value are estimated using the ratios between the weight of products leaving an enterprise and the weight of inputs it used to create them. These values are provided in live weight equivalents (e.g., the weight of fish in the can) and not the total weights (e.g., weight of cans including fish, liquids, and tin). Gains in value are estimated in a similar manner, using the ratios between the value of products and inputs flowing through each enterprise.

Calculations for estimating (i) revenue, (ii) profit, (iii) contributions to GDP, and (iv) employment for all enterprises described within value chain are expressed with the following equations,

$$L_c = W_{p,c} \cdot \prod_{e=1}^{c} (\frac{W_{i,e}}{W_{p,c}}) \tag{1}$$

where L_c is the live weight equivalent for a given value enterprise for which the value chain holds enterprises from the first (a producer) to the last element in the chain (c), $W_{p,e}$ is the weight of products for the enterprise (e), and $W_{i,e}$ is the weight of input ("raw material") for the same enterprise.

$$R_p = W_p \cdot (R_a + R_e + R_i + R_s) \tag{2}$$

where R_p is the overall production revenue for the enterprise. Revenues from subsidies (U) are calculated from,

$$U = W_p(U_e + U_o) \tag{3}$$

$$\text{Total revenue } (R) = R_p + U \tag{4}$$

$$\text{Cost of input and operation } (I) = W_p(I_c + I_e + I_i + I_s + C_m + C_l + C_c) \tag{5}$$

$$\text{Cost of observers } (O) = W_p(C_o \cdot O_r) \tag{6}$$

$$\text{Taxation costs } (T) = W_p(T_e + T_x + T_p + T_v + T_i + T_l) \tag{7}$$

$$P_w = \begin{cases} W_p \cdot (P_s + P_h), & \text{if using a wage system} \\ W_p \cdot V_{f,s}(S_s + S_h), & \text{if using a share system} \end{cases} \tag{8}$$

where $V_{f,s}$ is the value of the product (by fleet and by species) per unit weight.

$$P_o = \begin{cases} W_p \cdot (P_f + P_m), & \text{if using a wage system} \\ W_p \cdot V_{f,s}(S_f + S_m), & \text{if using a share system} \end{cases} \tag{9}$$

$$\text{Total costs } (C) = I + O + T + P_w + P_S \tag{10}$$

We calculate the number of jobs for workers (J_w) and owners (J_o), and the total number of jobs from the sum of J_w and J_o.

$$J_w = W_p(J_s + J_h) \tag{11}$$

$$J_o = W_p(J_f + J_m) \tag{12}$$

Further the numbers of dependents of workers (D_w) and owners (D_o) is calculated from

$$D_w = W_p(D_s \cdot J_s + D_h \cdot J + h) \tag{13}$$

$$D_o = W_p(D_f \cdot J_f + D_m \cdot J_m) \tag{14}$$

which can be summed to give the total number of dependents, $D = D_w + D_o$.

For producers, it is assumed that the number of jobs is proportional to effort, while their income depends on the catch value of the catches.

The socio-economic indicators (i.e., i-iv, see above) can be summarized by: (a) functional group, (b) fishing fleet, and (c) any enterprise within the value chain, or (d) for the whole fisheries sector. Moreover, the contributions to the GDP and employment can be divided between activities taking place at sea and on land to estimate income and employment multipliers (e.g., how many jobs (or $) are made on land for each job (or $) made at sea) by fleet, functional groups and across the fisheries system.

At the Ecopath stage, the value chain models can be used to address multiple questions related to characterizing the economic network. For example: Which fishing fleets are the most important contributors to national employment (at sea and on-land)? Are the income multipliers similar among functional groups and fishing fleets? How big is the fishing industry in economic terms? Do mackerels contribute more to a country's GDP when fished by purse seiners or by gillnetters? Are canneries and seafood freezing plants paying similar wages to the women they employ? How many people are employed in export-driven seafood supply chains in comparison to those selling locally? Are people earning annual salaries above the minimum wage across all enterprises in the seafood value chain?[12] [13] [14]

Value chains are good tools for highlight the roles played by marginalized groups within the fisheries sector (*e.g.*, women, small-scale fishers, subsistence fishers) in a systematic manner. For example, characterizing seafood value chains in Peru under this approach revealed that supply chains starting with small-scale fishers were the main contributors to employment and GDP across the fisheries sector, although being responsible for only 15% of the country's catch.[15] Additionally, the implementation of this approach in Baía Formosa (Brazil) allowed the users to quantify the indirect contribution of subsistence fisheries to local economies.[16]

Furthermore, understanding value chains' flows and structures opens new opportunities for addressing "what-if" questions regarding the end use of marine living resources, and how these affect their potential contribution to the economy, employment and job security (a known limitation of input-output models)[17]. For example, a study revealed that a transition from the fishmeal-dominated *status quo* to a hypothetical scenario where all anchoveta (*Engraulis ringens*) landed in Peru were used for canning; would result in a 53% reduction in the country's fishmeal production, a 21% increase in fisheries sectors' profitability, a 183% increase in job creation, and 179 times more seafood production.[18] Moreover, national and provincial value chains can be compared to highlight how the same functional groups and fishing fleets can have dissimilar employment and income multipliers at different spatial scales (revealing the local importance of certain functional groups or fishing fleets).[19]

At the Ecosim model stage, value chain models can be used for equilibrium analyses. For example, Christensen et al.[20] sought to assess the maximum sustainable yield (MSY) by setting a constant fishing effort over a 25 year-long Ecosim run, letting the system reach a steady state and then repeating the run with a new fishing effort level. They explored a wide range of effort levels (from no exploitation to overexploitation) on a theoretical fleet targeting tuna. In each step revenue, fishing costs, income, and employment for the fleets and the entire supply chain, were registered. This approach allowed researchers to: (i) highlight trade-offs between fishing fleets (e.g., tuna fleet vs mackerel fleet), (ii) showcase how fishing costs are non-linear (although commonly assumed to be so in equilibrium analysis [6]), and (iii) highlight the strengths of using *MSY* instead of the Maximum Economic Yield (MEY) as a descriptor for the maximum socio-economic benefits that can be achieved by a fishery.[21]

Moreover, value chain models can also be used together with Ecosim and Ecospace for studies involving hind- and forecasts. This tool is robust for expressing the socio-economic outcomes of "what-if" scenarios (from climate change to fisheries policies), and for testing the effects of broader economic policies (e.g., the introduction of new taxes or the elimination of fuel subsidies). Moreover, granted that it allows users to define the cost-income structure of the fleets in a more detailed manner (*Ecopath > Output > Tools > Value chain > Components > Producers*), even if only producer data is available, the Value Chain plug-in can strengthen estimates for the fishing profitability over time, as well as simulate the evolution of particular variable cost (e.g., fuel costs) across scenarios.

Yet, it is important to note that in these instances, food web modelling outputs (i.e., catches per fleet per functional group) will enter the value chain model without affecting its base parametrization. Thus, some value chain outputs might be misleading given that value chain parameters are not equally stable over time. For example, processing yields (e.g., the difference between the total weight of tuna entering a cannery and the weight of tuna in the cans coming out of it) tend to be quite stable over time, unless important changes in processing technology come to play. Alternatively, if a processing industry (e.g., reduction industry) is growing in a country, then the number of workers or owners per tonne of processed fish will vary substantially between years. Thus, if the processing capacity is in excess in the year corresponding to the base value chain model, then projections on the contribution this industry for total employment might be overestimated in future years (as the processing plant could still take more fish without having to adjust its labour force).

Given this issue, it is key to ground model outputs with data when using the value chain models, particularly for medium to long-term projections. First and foremost, users should highlight, when reporting results, that uncertainty in model outputs increases substantially over time. The classic economic assumptions of *ceteris paribus* will be damaging to fisheries system if the resulting management advice is given based exclusively on socio-economic indicators using a single set of value chain parameters. Just as one should avoid giving fisheries management advice using only Ecopath models, one should also avoid using static value chains. Solutions for these limitations can be undertaken by:

1. Developing multiple value chain models for consecutive years (e.g., 5 or 10 years) and running Ecosim over each value chain parametrization. The outputs could be plotted together to highlight the consequences of parameter and structural uncertainty in the socio-economic projections, or

2. Constructing a base value chain model for a 5-year or 10-year period (by averaging annual parameter values over time) and then using the coupled value

chain-Ecosim runs only to study changes over that same period or a future period of the same length.

The latter solution is certainly a less preferable one. However, access to data might make it difficult to develop multiple consecutive full value chain models. Notwithstanding, adopting a scenario approach to value chain outputs is still useful, particularly by forcing users to explicitly state their hypothesis on what is assumed to remain constant over time and why.

Nonetheless, EwE's Value Chain plug-in grants unique capabilities for directly estimating the income and employment multipliers of specific marine living resources (i.e., through EwE functional groups), producers (i.e., fishing fleets) and the fisheries sector. Moreover, it provides modelers with a platform to synthesize large amounts of socio-economic knowledge of fisheries systems to provide a description of its whole economic subsystem in a succinct but comprehensive manner, much like what an Ecopath model does for an ecosystem's food web.

The process of constructing a value chain model requires developing a working hypothesis of its structure and understanding what information is available, how it is stored, who has access to it, and who collects and updates it. This process is quick to reveal areas (e.g., enterprises, items within their cost structure or segments of the value chain) with information deficits, that can be used to prioritize research and monitoring efforts. Moreover, if conducted in a participatory and inclusive manner, this process can help strengthen seafood traceability,[22] improve managers' understanding of leverage points along the supply chain, and be used to simulate the consequences of potential governmental interventions (both at sea and on land) on the national economy. For example, value chain models in EwE can be used together with the Management Strategy Evaluation (MSE) Module or the CEFAS MSE plug-in (See the MSE CEFAS tutorial)[23] to simulate the effects of implementing alternative management procedures (e.g., harvest control rules) on individual or multiple functional groups using indicators that describe the ecological, economic and social components of fisheries systems. This allows modelers the capacity to directly quantify trade-offs amongst management objectives and harvest strategies, while highlighting how the costs and benefits of the different management procedures are distributed among stakeholders and their enterprises within the system. [24] [25] [26].

Notes

1. Roy N., R. Arnason, W.E. Schrank, The identification of economic base industries, with an application to the Newfoundland fishing industry, Land Economics 85 (2009) 675–691. http://le.uwpress.org/content/85/4/675.short

2. Goodwin N., J.M. Harris, J.A. Nelson, P.J. Rajkarnikar, B. Roach, M. Torras, Microeconomics in context, 4th ed., Routledge, 2019

3. The Gross Domestic Product (GDP) is a monetary measure of the market value of all the final goods and services produced in an economy within a year. For more information see:

https://data.oecd.org/gdp/gross-domestic-product-gdp.htm

4. Dyck A.J., U.R. Sumaila, Economic impact of ocean fish populations in the global fishery, Journal of Bioeconomics 12 (2010) 227–243. https://doi.org/10.1007/s10818-010-9088-3

5. Dyck A.J., U.R. Sumaila, (2010) *op. cit.* https://doi.org/10.1007/s10818-010-9088-3

6. Christensen V., J. Steenbeek, P. Failler, A combined ecosystem and value chain modeling approach for evaluating societal cost and benefit of fishing, Ecological Modelling 222 (2011) 857–864. https://doi.org/10.1016/j.ecolmodel.2010.09.030

7. Christensen V., et al. (2011) op.cit. https://doi.org/10.1016/j.ecolmodel.2010.09.030

8. Christensen V., S. De la Puente, J.C. Sueiro, J. Steenbeek, P. Majluf, Valuing seafood: The Peruvian fisheries sector, Mar Policy 44 (2014) 302–311. https://doi.org/10.1016/j.marpol.2013.09.022

9. Gozzer-Wuest R., J.C. Sueiro, J. Grillo-Núñez, S. De la Puente, M. Correa, T. Mendo, J. Mendo, Desafiando la tradición de país harinero: Una mirada económica de la actividad pesquera de Piura, Perú, Mar Fish Sci Mafis 35 (2022) 255–274. https://doi.org/10.47193/mafis.3522022010507

10. Bevilacqua A.H.V., R. Angelini, J. Steenbeek, V. Christensen, A.R. Carvalho, Following the Fish: The Role of Subsistence in a Fish-based Value Chain, Ecological Economics 159 (2019) 326–334. https://doi.org/10.1016/j.ecolecon.2019.02.004

11. Christensen V., et al. (2011) op.cit. https://doi.org/10.1016/j.ecolmodel.2010.09.030

12. Christensen V et al. (2014) *op.cit.* https://doi.org/10.1016/j.marpol.2013.09.022

13. Gozzer-Wuest R et al. (2022) https://doi.org/10.47193/mafis.3522022010507

14. Bevilacqua et al. *op.cit* (2019). https://doi.org/10.1016/j.ecolecon.2019.02.004

15. Christensen V et al. (2014) *op.cit.* https://doi.org/10.1016/j.marpol.2013.09.022

16. Bevilacqua et al. *op.cit* (2019). https://doi.org/10.1016/j.ecolecon.2019.02.004

17. Seung C.K., E.C. Waters, A Review of Regional Economic Models for Fisheries Management in the U.S., Marine Resource Economics 21 (2006) 101–124 https://www.jstor.org/stable/42629497

18. Majluf P.Y., S. De la Puente, V. Christensen, The little fish that can feed the world, Fish and Fisheries 18 (2017) 772–777. https://doi.org/10.1111/faf.12206

19. Gozzer-Wuest R et al. (2022) https://doi.org/10.47193/mafis.3522022010507

20. Christensen V., et al. (2011) op.cit. https://doi.org/10.1016/j.ecolmodel.2010.09.030

21. Christensen V., MEY = MSY, Fish and Fisheries 11 (2010) 105–110. https://doi.org/10.1111/j.1467-2979.2009.00341.x

22. Fox M., M. Mitchell, M. Dean, C. Elliott, K. Campbell, The seafood supply chain from a fraudulent perspective, Food Secur. 10 (2018) 939–963. https://doi.org/10.1007/s12571-018-0826-z

23. Mackinson S., M. Platts, C. Garcia, C. Lynam, Evaluating the fishery and ecological consequences of the proposed North Sea multi-annual plan, PLoS ONE 13 (2018) e0190015-23. https://doi.org/10.1371/journal.pone.0190015

24. Christensen V., et al. (2011) op.cit. https://doi.org/10.1016/j.ecolmodel.2010.09.030

25. Nielsen J.R. et al. (2018) *op. cit.* https://doi.org/10.1111/faf.12232

26. Steenbeek J., J. Buszowski, V. Christensen, E. Akoglu, K. Aydin, N. Ellis, D. Felinto, J. Guitton, S. Lucey, K. Kearney, S. Mackinson, M. Pan, M. Platts, C.J. Walters, Ecopath with

Ecosim as a model-building toolbox: Source code capabilities, extensions, and variations, Ecological Modelling 319 (2016) 178–189. https://doi.org/10.1016/j.ecolmodel.2015.06.031

42.

Price elasticity

Seafood prices vary, often in unclear ways. Maybe a rare species doesn't have much of a market because people prefer to eat what they know – what they know how to prepare and like. Or, the rarer a species gets, the higher price it fetches if demand keeps steady. Abalone and bluefin tuna might be examples. Demand for seafood can also impact supply where increased demand leads to higher prices, which in turn may drive more fishers to target the species in question, or *vice versa* increased supply may lead to lower landings prices, especially where processing capacity is a limiting factor.

So, landings prices are not constant, but may change due to demand and supply factors. How do we consider that in EwE? The starting point is the landings prices in the Ecopath input parameters. Almost all EwE applications have simply taken those landings prices as being constant, and evaluated management options based on that assumption (corresponding to the solid lines in Figure 1). But landings prices are not constant! Economists deal with that issue through what is called price elasticity[1].

For economists, price elasticity (e_p) is generally defined as,

$$e_p = \frac{dQ/Q}{dP/P} \tag{1}$$

where Q is the quantity demanded and P is the price. So, price elasticity represents the ratio between how much demand changes and how much the price changes. If prices increase, demand will decrease is the general expectation for a good with price elasticity. If for instance a 10% increase in price leads to a 10% decrease in demand, the situation is called "unit elasticity", and it results in revenue being constant (corresponding to the stippled lines in Figure 1).

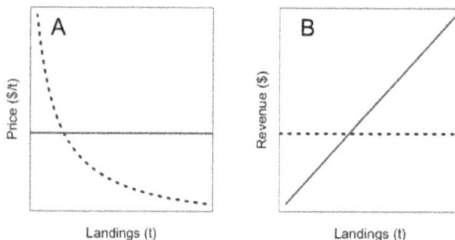

Figure 1. A. Two examples of price elasticity. The dotted line represents unit elasticity where an increase in landings of x% leads to a corresponding decrease in landing price of x%. The horizontal line represents a perfectly inelastic situation where price is inde-

pendent of supply, i.e. the default assumption in EwE where price is independent of landings (if not including price elasticity). B. Revenue (total landing value) for the situation with unit elasticity (dotted line) and inelastic (straight line) where revenue is proportional to landings.

The situation is a bit more complicated for capture fisheries where there often is a complex relationship between demand and supply – and for that matter with production capacity where more boats doesn't automatically result in more supply. For "traditional" seafood species prices may be relatively stable as landings change, perhaps with demand increasing prices with low supply, and with dropping prices when processing capacity becomes a limiting factor. For less traditional seafood, which may be the rarer species that consumers are not used and willing to purchase, prices may be low with low supply and only increase if and when catches increase.

In EwE, we consider the complex pricing pattern through a price elasticity functionality in Ecosim. In Ecosim one can define the relationship between supply and resulting landing price. The supply may be for one or more functional groups caught by a one or more fishing gears, and the resulting landing price may then be changed proportionally for groups – gear combinations that are specified separately. So, in principle one could have that the supply of one species impacts the price of another. Take as a hypothetical example that increase landings of walleye pollock for surimi (imitation crab) production might impact the landing price for king crab.

The price elasticity functionality while defined in Ecosim can subsequently be used anywhere in EwE where landed value is calculated, e.g., in the value chain, policy optimization, or MSE.

For hands-on details of how price elasticity is implemented in EwE, please see the Anchovy Bay price elasticity tutorial and the EwE User Guide.

Notes

1. For more about price elasticity, see, e.g., the open textbook Principles of Microeconomics by Emma Hutchinson, University of Victoria.

43.

Fishing policy exploration

Santiago de la Puente

Ecosim contains a formal optimization routine (*i.e.*, an "open loop" simulation[1]) that allows users to search for fisheries policies that would maximize long-term management goals.[2] The routine uses a multi-criterion objective function[3] representing five common fisheries management goals:

- **Maximize fisheries rent** (net economic value, R): Where profits are calculated as a function of the value of the catch minus the cost of fishing.

- **Maximize fisheries social benefits** (J): Where social benefits are expressed as the employment supported by the fleet, such that the number of jobs per ton of fish caught per fleet are proportional to fishing effort.

- **Maximize mandated rebuilding** of a functional group (B_{lim}): Where rebuilding targets are set by describing a threshold biomass (relative to the biomass on the base Ecopath model) for a functional group whose biomass is low or declining.

- **Maximize species diversity** *(D):* Where diversity is approximated using Kempton's Q75 index.

- **Maximize ecosystem structure** or "health" *(B/P):* Where average longevity across functional groups is regarded as a proxy for ecosystem maturity. Thus, ecosystem configurations that favour higher biomasses for groups with low Production/Biomass ratios are regarded as more desirable.

This routine works by affecting relative fishing effort levels *(E)* by fleet type *(fl)*. Using a nonlinear optimization procedure, it seeks to iteratively improve the objective function by changing relative fishing rates (by producing time series of relative fleet sizes).[4]

$$f(E_{fl}) = \text{Max}(w_1 R + w_2 J + \left\{ \begin{array}{l} W_3 \cdot (B_{lim} - B), \text{ if } B < B_{lim} \\ 0, \text{ if } B \geq B_{lim} \end{array} \right\} + w_4 D + w_5 \frac{B}{P}) \qquad (1)$$

Ecosim uses relative fleet sizes to calculate fishing mortality rates by fleet type. The basic assumption is that the mix of fishing rates over functional groups remains constant for each fleet type. However, it is also possible to account for hyperstability and hyperdepletion in EwE[5] so that, for example, some functional groups might retain high fishing rates even at lower levels of biomass.

The optimization routine used by EwE allows for maximizing economic objectives under scenarios of full cooperation (i.e., all incomes and costs are pooled, and profits are shared among all fishers and across fleets), constrained cooperation (i.e., maximizing profits across all fleets but where each fleet has to remain economically viable on its own), and full competition (i.e., treating each fleet as a separate economic entity, and seeking to maximize fleet-specific rent).[6][7][8] [39–41]. Additionally, it allows users to explore changes in trade-off schedules by: (i) varying discount rates in the net present value calculation,[9][10] and (ii) incorporating data from the value chain plugin (*i.e.*, fleet level vs supply chain level consequences in terms of net present value and jobs).

Finally, there is an alternative search procedure for optimum fishing patterns that maximize a logarithm-based portfolio utility function. When applied, this portfolio utility function embodies a risk-adverse objective function, as its logarithmic configurations heavily penalizes low values (e.g., years with low economic rent or contributions to employment).

Explore this tool further through the Trade-offs between policy objectives tutorial.

Notes

1. The control action is independent of the output of the system (no option for feedback between system outputs and inputs).

2. Christensen V, C.J. Walters, Trade-offs in ecosystem-scale optimization of fisheries management policies, Bulletin of Marine Science 74 (2004) 549–562.

3. A weighted sum of social, economic, and ecological indicators. It is important to note that allocating different weights (w) to each indicator type in equation might heighten conflicts or make trade-offs more explicit. Thus, analyzing alternative weighing schemes within the multi-criterion objective function is a topic worth exploring

4. Christensen V, C.J. Walters (2004) *op.cit.* Bull. Mar. Sci.

5. Hyperstability and hyperdepletion can be incorporated via de density-dependant catchability parameter in Ecosim's "Group info" form (*Ecosim > input > Group info*).

6. Christensen V, C.J. Walters (2004) *op.cit.* Bull. Mar. Sci.

7. Araújo J.N., S. Mackinson, R.J. Stanford, P.J.B. Hart, Exploring fisheries strategies for the western English Channel using an ecosystem model, Ecological Modelling 210 (2008) 465–477. https://doi.org/10.1016/j.ecolmodel.2007.08.015

8. Heymans J.J., U.R. Sumaila, V. Christensen, Policy options for the northern Benguela ecosystem using a multispecies, multifleet ecosystem model, Progress in Oceanography 83 (2009) 1–9. https://doi.org/10.1016/j.pocean.2009.07.013

9. Sumaila U.R., C.J. Walters, Intergenerational discounting: a new intuitive approach, Ecological Economics 52 (2005) 135–142. https://doi.org/10.1016/j.ecolecon.2003.11.012

10. Dichmont C.M., N. Ellis, R.H. Bustamante, R. Deng, S. Tickell, R. Pascual, H. Lozano-Montes, S. Griffiths, Evaluating marine spatial closures with conflicting fisheries and conservation objectives, J. Appl. Ecol. 50 (2013) 1060–1070. https://doi.org/10.1111/1365-2664.12110.

44.

External bio-economic models

Santiago de la Puente

No model is capable of doing everything, nor should it be. However, EwE models can be linked or coupled with external routines forming an ecological-social-economic modelling chain. For example, some authors have coupled Ecosim with bioeconomic models to get a better handle on fishing effort dynamics. For each time step, the bioeconomic model uses the predicted catches per fleet (Ecosim output) to estimate the fleets profitability and predict the next time steps' fishing effort level, which are then reintroduced as inputs for Ecosim.[1][2][3]

In other cases, EwE models have been linked with Input-Output (I-O) models,[4] Social Accounting Matrices (SAM)[5] and Computable general equilibrium (CGE) models.[6] These tools also allow estimation of the broader economic impact of the fisheries across national or regional economies (much like the Value Chain plug-in). I-O models characterize the flows of goods in a symmetrical industry by industry format (i.e., goods supplied vs consumed) and are used to estimate direct (measures of actual expenditures by establishments operating in the sector), indirect (measures of economic activity of other industries supplying an industry or using its outputs) and induced economic effects of a particular industry (measures of economic impact derived from the expenditure of salaries gained in the sector on other sectors of the economy).[7] SAMs are extensions of input-output models, with their main advantage being that they consider the social-economic linkages as well as other transactions (such as linkages between production and household sectors).[8] Finally, CGE models provide an analytical framework to assess the impact of fishery policies on regional economies and social welfare.[9]

Notes

1. Dichmont C.M., N. Ellis, R.H. Bustamante, R. Deng, S. Tickell, R. Pascual, H. Lozano-Montes, S. Griffiths, Evaluating marine spatial closures with conflicting fisheries and conservation objectives, J. Appl. Ecol. 50 (2013) 1060–1070. https://doi.org/10.1111/1365-2664.12110

2. Lee K., J. Apriesnig, H. Zhang, Socio-Ecological Outcomes of Single-Species Fisheries Management: The Case of Yellow Perch in Lake Erie, Front. Ecol. Evol. 9 (2021) 703813. https://doi.org/10.3389/fevo.2021.703813

3. Apriesnig J.L., T.W. Warziniack, D.C. Finnoff, H. Zhang, K.D. Lee, D.M. Mason, E.S. Rutherford, The consequences of misrepresenting feedbacks in coupled human and environmental models, Ecol. Econ. 195 (2022) 107355. https://doi.org/10.1016/j.ecolecon.2022.107355

4. Byron C.J., D. Jin, T.M. Dalton, An Integrated ecological–economic modeling framework for the sustainable management of oyster farming, AQC 447 (2015) 15–22. https://doi.org/10.1016/j.aquaculture.2014.08.030

5. Wang Y., J. Hu, H. Pan, S. Li, P. Failler, An integrated model for marine fishery management in the Pearl River Estuary: Linking socio-economic systems and ecosystems, Marine Policy 64 (2016) 135–147. https://doi.org/10.1016/j.marpol.2015.11.014

6. Wang Y., J. Hu, H. Pan, P. Failler, Ecosystem-based fisheries management in the Pearl River Delta: Applying a computable general equilibrium model, Marine Policy 112 (2020) 103784. https://doi.org/10.1016/j.marpol.2019.103784

7. Byron et al. (2015) *op. cit.* https://doi.org/10.1016/j.aquaculture.2014.08.030

8. Wang et al. (2016) *op.cit.* https://doi.org/10.1016/j.marpol.2015.11.014

9. Wang et al. (2020) *op. cit.* https://doi.org/10.1016/j.marpol.2019.103784

VII

Ecospace Introduction

45.

Introduction to Ecospace

Setting the stage

Marine ecosystems are complex systems affected by the state of the environment and a myriad of human activities. The full impact of these human activities on the complex ecosystem are becoming more critical to understand, and are now requested by agencies concerned with numerous policies and strategies that involve spatial management actions (in Europe, for instance by the Common Fisheries Policy, Marine Strategy Framework Directive, Farm to Fork, Zero Pollution and Biodiversity Strategies). All of these have divergent end-points, but all require an understanding of how multiple forms of human activities impact marine ecosystems. To better manage our impact on marine ecosystems, and notably consider trade-offs in management, there is a need to advance scientific capabilities to provide both quantitative descriptions and quantitative evaluations of the effect of spatial management interventions, factoring in plausible future changes in climate and human activities.

Spatial-temporal ecosystem modeling has grown in capacity, complexity, and focus to undertake such complex tasks, and is increasingly considered an indispensable tool to contribute to policy and management, including multi-sectoral Ecosystem Based Management (EBM) and Marine Spatial Planning (MSP).

Among the tools available, the temporal and spatial dynamic model EwE approach is of special interest for its wide range of applications across ecosystem types. *Ecospace* has notably been used to contribute to EBM, most often to a subset of EBM, Ecosystem-Based Fisheries Management, or EBFM. Applications of *Ecospace* include the evaluation of spatial trophic interaction patterns, modeling of species distribution based on habitat suitability, the assessment of Marine Protected Area (MPA) placement and connectivity, harvest allocations and, more recently, environmental impact analysis and the assessment of episodic mortality events, effects of changes in nutrient inputs, climate change, and cumulative impacts (e.g., [1,2,3,4,5,6,7,8,9,10,11,12,13,14,15,16,17,18].

The development of Ecospace has involved an evolutionary process where many additional capabilities have been developed over the years in response to requests by users. The more recent advancements include the Habitat Foraging Capacity Model (HFC[19])and the Spatial Temporal Data Framework (STDF[20]), which have enabled Ecospace applications to fully consider climate variability and change, taking into account different types of uncertainty[21,22]. These innovations and a substantial increase in the various applications of EwE[23], provided a call for an update of the Ecospace, which led to a book chapter by de Mutsert et al.[24] upon which a number of the chapters in the present text book are based (as attributed). The present text book and the EwE User Guide seek to present the most up-to-date guidelines for using and understanding Ecospace's capabilities and challenges.

About *Ecospace*

The Ecospace model is a spatially explicit time dynamic model based on the *Ecopath* mass-balance and *Ecosim* time dynamic routines[25][26]. It applies the same set of differential equations as used in Ecosim, executed for each functional group and cell in a grid of cells. In Ecosim, a set of differential equations is defined based on the biomass components of change for consumer functional groups, expressed as

$$\frac{dB_i}{dt} = g_i \cdot \sum_{j=1}^{n} Q_{ji} - \sum_{j=1}^{n} Q_{ij} + I_i - (F_{it} + e_i + M0_{it}) \cdot B_{it} \tag{1}$$

where B_{it} is the biomass of i at time t, g_i is the growth efficiency, I_i is the immigration rate; F_{it} is the mortality rate due to harvesting (fishing mortality); e_i is the emigration rate; and $M0_i$ the other mortality (mortality not explained in the model). The terms Q_{ji} and Q_{ij} represent the consumption due to predation by j on i, and by i on j, respectively. For primary producers, the consumption rate term is replaced by a production rate function $f(B_{it})$) represents primary production rate as a function of the group biomass[27]; that function is nonlinear, representing competition effects for light and nutrients.

The consumption rates Q_{ji} are predicted using equations from foraging arena theory (see chapter), where the biomass of prey i is split between a vulnerable (V_{ij}) and a non-vulnerable (B_i-V_{ij}) component. The vulnerability exchange parameters used in predicting the various Q_{ji} mainly represent spatial distribution and movement behaviors at very fine scales, typically far smaller than the size of Ecospace spatial grid cells.

In Ecospace, the spatial extent of the ecosystem is represented by a grid of cells, each of which can be defined as land or water, and each cell can have characteristics or attributes like a habitat type. Ecospace then represents the biomass (B) and consumption (Q) dynamics over a two-dimensional space as well as time[28]. Space, time, and state are considered discrete variables by using the Eulerian approach, which treats movements as "flow" of organisms among fixed spatial reference cells.

In the original Ecospace model[29], a first step of parameterizing entailed the definition of a base map based on habitat information (depth strata, bottom type, etc.) in the study area. Species preferences were then (and still can be if more elaborate spatio-temporal habitat use functions are not used) assigned to these habitat types based on the biology and ecology of the species included in each functional group of the ecosystem model, their depth distributions, their preferred sediment type, etc. In addition, the original *Ecospace* model required for habitat definitions,

    1. the dispersal (spatial mixing) rate of each functional group in "preferred" habi-

tats,

2. the relative dispersal rate in "non-preferred" habitats, and

3. the relative feeding rate in non-preferred habitat by functional group.

Fishing mortality rate for each cell can represent effects of fishing effort by multiple fishing fleets, and each fishing fleet can be depicted as operating in a specific region and habitat type, and cells can be defined as protected areas for particular or all fishing fleets. Fishing effort is assumed to move between grid cells over time in response to spatial and temporal variation in profitability of fishing.

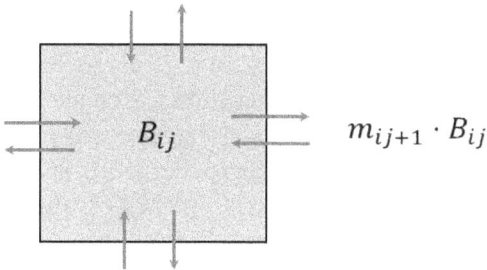

Figure 1. For each cell, the inbound dispersal rate I_i is the sum of emigration flows from the four surrounding cells, while the outbound instantaneous dispersal rates m_i from a given cell in Ecospace make up the basic Ecosim emigration rates e_i, and vary based on the pool type, cell conditions/habitat, and response of organisms to predation risk and feeding conditions.

Moreover, spatial variations of primary productivity and fishing costs can be defined as initial conditions for the basic model.

For trophic interactions, fishing, and movement calculations, biomass is considered as homogeneous within each cell and movement of biomass and flows is allowed across the borders to adjacent cells. For each cell, the immigration rate I_i of Eq. 1 is assumed to consist of up to four emigration flows from the surrounding cells (Figure 1). The emigration flows ($B_{out,rci}$) are in turn similarly represented by instantaneous movement rates m_i times the biomass density in the cell (B_{rci}) with the sum of those loss rates like $m_{i,j+1}$ representing the emigration rate e_i in eq. 1:

$$B_{out,rci} = \sum_{d=1}^{4} m_{id} \cdot B_{rci} \qquad (3)$$

where (*rci*) represents cell row and column for group *i*, and *d* is movement direction (up, down, left or right).

The instantaneous emigration rates $m_{i,d}$ from a given cell in Ecospace are assumed to vary based on the functional group, habitat preferences, and can be set to vary with trophic conditions within each cell (to model responses of organisms to predation risk and feeding conditions). The probability of movement of organisms towards favourable habitats was in the original Ecospace formulation calculated by means of a "habitat gradient function" for each mapped habitat type and species or group i. Biomass dynamics in unsuitable cells were modified by predicting higher rates of emigration, lower feeding rates, and/or higher vulnerability to predation, and a complex gradient calculation continues to be used so as to modify dispersal rates to cause higher movement rates of biomass toward suitable cells.

In more recent versions of Ecospace, a habitat capacity model has been included to estimate cell-specific continuous habitat suitability factors where the area that species can feed in each cell is determined by functional responses to multiple environmental factors[30]. See the habitat capacity chapter. It is optional whether to use a habitat and/or habitat suitability for any given group, though in many recent applications habitat suitability is used predominantly while habitats mainly are used for defining where fleets can operate.

The very large equation system represented by eq. 1 with mixing terms is solved numerically on one month time steps using an implicit integration method (BDF2, second order backward differentiation), that works very well for long-lived species but tends to predict short lived species to change more slowly than Ecosim predicts. As in Ecosim, the Ecospace BDF integration does not "see" the very rapid boom-bust dynamics that can be exhibited by groups with high P/B (e.g. phytoplankton and short-lived zooplankters), but instead predicts an average biomass for these groups.

For advanced applications involving multi-stanza groups, there are two Ecospace option[31]. The first "multi-stanza" solution option. is to keep track of overall multi-stanza numbers at age over the map while predicting local variation in abundance from concentration patterns predicted from spatial m_{ij} variations of the differential equation system. The second or "IBM" option is to divide the multi-stanza recruitment numbers at each time step into a large number of packets of individuals, then simulate random and directed movements of these packets over the map as the organisms grow. Details of the IBM equations are presented later in the chapter on Spatial implementation of multi-stanza and IBM.

Attribution The first section of this chapter is based on de Mutsert K, Marta Coll, Jeroen Steenbeek, Cameron Ainsworth, Joe Buszowski, David Chagaris, Villy Christensen, Sheila J.J. Heymans, Kristy A. Lewis, Simone Libralato, Greig Oldford, Chiara Piroddi, Giovanni Romagnoni, Natalia Serpetti, Michael Spence, Carl Walters. 2023. Advances in spatial-temporal coastal and marine ecosystem modeling using Ecopath with Ecosim and Ecospace. Treatise on Estuarine and Coastal Science, 2nd Edition. Elsevier. https://doi.org/10.1016/B978-0-323-90798-9.00035-4, adapted with permission, License Number 5651431253138.

The second section of the chapter is partly based on Christensen, V, M Coll, J Steenbeek, J Buszowski, D Chagaris, and CJ Walters. 2014. Representing variable habitat quality in a spatial food web model. Ecosystems 17(8): 1397-1412. https://doi.org/10.1007/s10021-014-9803-3.

Rather than citing this chapter, please cite the sources

Notes

1. Alexander, K.A., Meyjes, S.A., Heymans, J.J., 2016. Spatial ecosystem modelling of marine renewable energy installations: Gauging the utility of Ecospace. Ecological Modelling, Ecopath 30 years – Modelling ecosystem dynamics: beyond boundaries with EwE 331, 115–128. https://doi.org/10.1016/j.ecolmodel.2016.01.016

2. Coll, M., Pennino, M.G., Steenbeek, J., Sole, J., Bellido, J.M., 2019. Predicting marine species distributions: Complementarity of food-web and Bayesian hierarchical modelling approaches. Ecological Modelling 405, 86–101. https://doi.org/10.1016/j.ecolmodel.2019.05.005

3. Dahood, A., de Mutsert, K., Watters, G.M., 2020. Evaluating Antarctic marine protected area scenarios using a dynamic food web model. Biological Conservation 251, 108766. https://doi.org/10.1016/j.biocon.2020.108766

4. De Mutsert, K., Lewis, K., Milroy, S., Buszowski, J., Steenbeek, J., 2017a. Using ecosystem modeling to evaluate trade-offs in coastal management: Effects of large-scale river diversions on fish and fisheries. Ecological Modelling 360, 14–26. https://doi.org/10.1016/j.ecolmodel.2017.06.029

5. De Mutsert, K., Lewis, K.A., White, E.D., Buszowski, J., 2021. End-to-End Modeling Reveals Species-Specific Effects of Large-Scale Coastal Restoration on Living Resources Facing Climate Change. Front. Mar. Sci. 8. https://doi.org/10.3389/fmars.2021.624532

6. Espinosa-Romero, M.J., Gregr, E.J., Walters, C., Christensen, V., Chan, K.M.A., 2011. Representing mediating effects and species reintroductions in Ecopath with Ecosim. Ecological Modelling 222, 1569–1579. https://doi.org/10.1016/j.ecolmodel.2011.02.008

7. Fouzai, N., Coll, M., Palomera, I., Santojanni, A., Arneri, E., Christensen, V., 2012. Fishing management scenarios to rebuild exploited resources and ecosystems of the Northern-Central Adriatic (Mediterranean Sea). Journal of Marine Systems 102–104, 39–51. https://doi.org/10.1016/j.jmarsys.2012.05.003

8. Libralato, S., Solidoro, C., 2009. Bridging biogeochemical and food web models for an End-to-End representation of marine ecosystem dynamics: The Venice lagoon case study. Ecological Modelling 220, 2960–2971. https://doi.org/10.1016/j.ecolmodel.2009.08.017

9. Martell, S.J.D., Essington, T.E., Lessard, B., Kitchell, J.F., Walters, C.J., Boggs, C.H., 2005. Interactions of productivity, predation risk, and fishing effort in the efficacy of marine protected areas for the central Pacific. Can. J. Fish. Aquat. Sci. 62, 1320–1336. https://doi.org/10.1139/f05-114

10. Masi, M.D., Ainsworth, C.H., Kaplan, I.C., Schirripa, M.J., 2018. Interspecific Interactions

May Influence Reef Fish Management Strategies in the Gulf of Mexico. Mar Coast Fish 10, 24–39. https://doi.org/10.1002/mcf2.10001

11. Okey, T.A., Banks, S., Born, A.F., Bustamante, R.H., Calvopiña, M., Edgar, G.J., Espinoza, E., Fariña, J.M., Garske, L.E., Reck, G.K., Salazar, S., Shepherd, S., Toral-Granda, V., Wallem, P., 2004. A trophic model of a Galápagos subtidal rocky reef for evaluating fisheries and conservation strategies. Ecological Modelling, Placing Fisheries in their Ecosystem Context 172, 383–401. https://doi.org/10.1016/j.ecolmodel.2003.09.019

12. Ortiz, M., Wolff, M., 2002. Spatially explicit trophic modelling of a harvested benthic ecosystem in Tongoy Bay (central northern Chile). Aquatic Conservation: Marine and Freshwater Ecosystems 12, 601–618. https://doi.org/10.1002/aqc.512

13. Piroddi, C., Akoglu, E., Andonegi, E., Bentley, J.W., Celić, I., Coll, M., Dimarchopoulou, D., Friedland, R., de Mutsert, K., Girardin, R., Garcia-Gorriz, E., Grizzetti, B., Hernvann, P.-Y., Heymans, J.J., Müller-Karulis, B., Libralato, S., Lynam, C.P., Macias, D., Miladinova, S., Moullec, F., Palialexis, A., Parn, O., Serpetti, N., Solidoro, C., Steenbeek, J., Stips, A., Tomczak, M.T., Travers-Trolet, M., Tsikliras, A.C., 2021. Effects of Nutrient Management Scenarios on Marine Food Webs: A Pan-European Assessment in Support of the Marine Strategy Framework Directive. Front. Mar. Sci. 8. https://doi.org/10.3389/fmars.2021.596797

14. Piroddi, C., Coll, M., Macias, D., Steenbeek, J., Garcia-Gorriz, E., Mannini, A., Vilas, D., Christensen, V., 2022. Modelling the Mediterranean Sea ecosystem at high spatial resolution to inform the ecosystem-based management in the region. Sci Rep 12, 19680. https://doi.org/10.1038/s41598-022-18017-x

15. Romagnoni, G., Mackinson, S., Hong, J., Eikeset, A.M., 2015. The Ecospace model applied to the North Sea: Evaluating spatial predictions with fish biomass and fishing effort data. Ecological Modelling 300, 50–60. https://doi.org/10.1016/j.ecolmodel.2014.12.016

16. Salomon, A.K., Waller, N.P., McIlhagga, C., Yung, R.L., Walters, C., 2002. Modeling the trophic effects of marine protected area zoning policies: A case study. Aquatic Ecology 36, 85–95. https://doi.org/10.1023/A:1013346622536

17. Vilas, D., 2022. Spatiotemporal Ecosystem Dynamics on the West Florida Shelf: Prediction, Validation, and Application to Red Tides and Stock Assessment. University of Florida.

18. Walters, C., Christensen, V., Walters, W., Rose, K., 2010. Representation of multistanza life histories in Ecospace models for spatial organization of ecosystem trophic interaction patterns. Bulletin of Marine Science 86, 439–459.

19. Christensen, V., Coll, M., Steenbeek, J., Buszowski, J., Chagaris, D., Walters, C.J., 2014. Representing Variable Habitat Quality in a Spatial Food Web Model. Ecosystems 17, 1397–1412. https://doi.org/10.1007/s10021-014-9803-3

20. Steenbeek, J., Coll, M., Gurney, L., Mélin, F., Hoepffner, N., Buszowski, J., Christensen, V., 2013. Bridging the gap between ecosystem modeling tools and geographic information systems: Driving a food web model with external spatial–temporal data. Ecological Modelling 263, 139–151. https://doi.org/10.1016/j.ecolmodel.2013.04.027

21. Coll, M., Steenbeek, J., 2017. Standardized ecological indicators to assess aquatic food webs: The ECOIND software plug-in for Ecopath with Ecosim models. Environmental Modelling & Software 89, 120–130. https://doi.org/10.1016/j.envsoft.2016.12.004

22. Steenbeek, J., Corrales, X., Platts, M., Coll, M., 2018. Ecosampler: A new approach to assessing parameter uncertainty in Ecopath with Ecosim. SoftwareX 7, 198–204. https://doi.org/10.1016/j.softx.2018.06.004

23. Colléter, M., Valls, A., Guitton, J., Gascuel, D., Pauly, D., Christensen, V., 2015. Global overview of the applications of the Ecopath with Ecosim modeling approach using the EcoBase models repository. Ecological Modelling 302, 42–53. https://doi.org/10.1016/j.ecolmodel.2015.01.025[/footnote] [footnote]Coll, M., Akoglu, E., Arreguín-Sánchez, F., Fulton, E.A., Gascuel, D., Heymans, J.J., Libralato, S., Mackinson, S., Palomera, I., Piroddi, C., Shannon, L.J., Steenbeek, J., Villasante, S., Christensen, V., 2015. Modelling dynamic ecosystems: venturing beyond boundaries with the Ecopath approach. Rev Fish Biol Fisheries 25, 413–424. https://doi.org/10.1007/s11160-015-9386-x

24. De Mutsert K, Marta Coll, Jeroen Steenbeek, Cameron Ainsworth, Joe Buszowski, David Chagaris, Villy Christensen, Sheila J.J. Heymans, Kristy A. Lewis, Simone Libralato, Greig Oldford, Chiara Piroddi, Giovanni Romagnoni, Natalia Serpetti, Michael Spence, Carl Walters. 2023. Advances in spatial-temporal coastal and marine ecosystem modeling using Ecopath with Ecosim and Ecospace. Treatise on Estuarine and Coastal Science, 2nd Edition. Elsevier. https://doi.org/10.1016/B978-0-323-90798-9.00035-4

25. Walters C, Christensen V, Pauly D. 1997. Structuring dynamic models of exploited ecosystems from trophic mass-balance assessments. Reviews in Fish Biology and Fisheries 7: 139-172. https://doi.org/10.1023/A:1018479526149

26. Christensen V, Walters C. 2004. Ecopath with Ecosim: methods, capabilities and limitations. Ecological Modelling 72: 109-139. https://doi.org/10.1016/j.ecolmodel.2003.09.003

27. Christensen and Walters. 2004. *op. cit.*

28. Walters C, Pauly D, Christensen V. 1999. Ecospace: prediction of mesoscale spatial patterns in trophic relationships of exploited ecosystems, with emphasis on the impacts of marine protected areas. Ecosystems 2: 539-554. https://doi.org/10.1007/s100219900101

29. Walters et al. 1999. *op. cit.*

30. Christensen, V., Coll, M., Steenbeek, J., Buszowski, J., Chagaris, D., Walters, C.J., 2014. Representing Variable Habitat Quality in a Spatial Food Web Model. Ecosystems 17, 1397–1412. https://doi.org/10.1007/s10021-014-9803-3

31. Walters, C., Christensen V, Walters W, Rose K. 2010. Representation of multi-stanza life histories in Ecospace models for spatial organization of ecosystem trophic interaction patterns. Bull. Mar. Sci. 86(2):439-459

46.

Habitat capacity

Villy Christensen; Marta Coll; Jeroen Steenbeek; Joe Buszowski; David Chagaris; and Carl J. Walters

Why are species where they are?

To bridge the gap between ecosystem models and species distribution models, the spatial-temporal explicit module Ecospace includes a habitat capacity model[1] that addresses the central question, "why are species where they are?" The overarching assumption in the habitat capacity model is that they are where they are because they prefer certain combinations of environmental and ecological conditions.

Prior to the inclusion of the habitat capacity model in Ecospace, species distribution models and ecosystem models offered limited capabilities to work jointly to produce needed integrated assessments: assessments that take both food web dynamics and spatial-temporal environmental variability into account. The habitat capacity model is fairly simple and its integration in EwE mainly implied replacing a binary habitat variable with a continuous habitat suitability factor, where the area that species can feed in each cell is determined by functional responses to multiple environmental factors. This modification builds on the fact that animal populations have lower local impacts as the size of their forage area increases. The habitat capacity model offers the ability to drive foraging capacity from multiple physical, oceanographic, and environmental factors such as depth, bottom type, temperature, salinity, oxygen concentrations, etc., which have cumulative impacts on the ability of functional groups to forage. Since cell capacity is calculated for every functional group at every time step, this modification makes the model fully temporal and spatially dynamic.

Using relative habitat capacity to predict spatial abundance

A reasonably simple and tractable way around the binary parameterization of habitat definition is to define a continuous relative habitat capacity C_{rcj} for each group j in each cell (with row and column) r,c, where C_{rcj} varies from 0.0 to 1.0, and is calculated for each cell as a function of a vector of habitat attributes $H_{rc} = (H1,H2,...Hv)_{rc}$ of that cell, i.e. $C_{rcj} = f_j(H_{rc})$. For example $H1$ might be water depth, $H2$ might be proportion hard bottom, $H3$ might be summer water temperature, etc. Figure 1 provides a schematic overview of the basic calculations in the habitat capacity model.

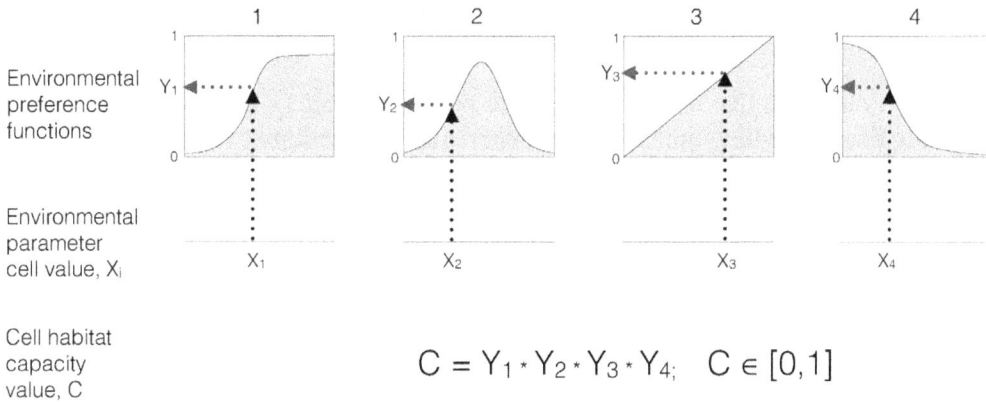

Figure 1. Schematic diagram of the habitat capacity model calculations with four (hypothetical) environmental preference functions (any number of functions is possible). During model run, cell-specific environmental parameter values can be read from data layers for each time step, and a cell-specific habitat capacity value is estimated as the product of the environmental preference values. No weighting is used, but weighting can be considered by altering the shapes of the environmental preference functions.

The proportion of a cell that a species (or functional group) can use is thus a continuous value from 0 to 1, and allows inclusion of as many environmental factors as needed to define the foraging capacity of a cell for a species in an Ecospace model.

If the functions $f_j()$ are chosen carefully, C_{rcj} can be updated over time with relatively little computational cost, for example by loading time-varying values of H_{rc} generated by other models or remotely sensed data for physical or biophysical change, and implemented using the spatial-temporal data framework of Ecospace, (see User Guide chapter for how-to).[2]

To include environmental factors with higher or lower weight, let the Y-axis values (Figure 1) vary more or less through the X-axis range.

In order to use the C_{rcj} habitat assessments, the C_{rcj} values have to be linked to trophic interaction dynamics to specify how C_{rcj} impacts food consumption and predation rates. A simple and reasonable way to represent this linkage is available through the basic foraging arena equations used to predict trophic interaction (food-web biomass flow) rates Eq. 1.

For this, the consumption rates Q_{ij} are based on the foraging arena theory (see chapter), where the biomass of prey i is split between a vulnerable (V_{ij}) and a non-vulnerable ($B_i - V_{ij}$) component. The transfer rate, called vulnerability (v_{ij}) between the two fractions determines the vulnerable biomass at time interval dt,

$$\frac{dV_{ij}}{dt} = v_{ij} \left(B_i - V_{ij} \right) - v_{ij} \, V_{ij} - \frac{a_{ij} \, V_{ij} \, B_j}{D_j} \tag{1}$$

where a_{ij} is the effective search rate for the predator j, and D_j represents loss of time searching due to handling time for the predator. The vulnerability parameter v_{ij} is a function of the maximum increase in predation mortality under the given predator/prey conditions (see vulnerability multiplier chapter). High values of v_{ij} imply large proportions of biomass (B_i) vulnerable to predator j (V_{ij}), and thus imply $V_{ij} = B_i$, and that the predator j is far from its carrying capacity with regards to prey i.

If we consider how *Ecosim* represents biomass dynamics (exclusive of spatial mixing effects), trophic interaction and fishery effects are modelled by equations of the basic form (looking at only one prey type to simplify the equation)

$$\frac{dB_j}{dt} = \frac{g_j \, a_{ij} \, v_{ij} \, B_j \, B_i}{2 \, v_{ij} + a_{ij} \, B_j} - Z_j \, B_j \tag{}$$

where Z_j is total instantaneous mortality rate of j, g_j is growth efficiency (corresponding to the production/consumption ratio, which can vary as predators grow in size), v_{ij} is prey vulnerability exchange rate, and a_{ij} is the rate of effective search by the predator. Note that in this model, vulnerable prey density V_{ij} is represented by the foraging arena equation Eq. 1), which simplified can be expressed when there is only one prey type i as

$$V_{ij} = \frac{v_{ij} \, B_i}{2 \, v_{ij} + a_{ij} \, B_j} \tag{3}$$

where predation pressure in a cell depends on the foraging arena area in that cell. If we assume that variation in relative habitat capacity for the predator means variation in the foraging arena area over which a species can forage successfully, we can include variation in relative habitat capacity in the model by dividing the denominator $a_{ij} \cdot B_j$ term by relative habitat size or capacity C_{rcj}, i.e.,

$$V_{ij} = \frac{v_{ij} \, B_i}{2 \, v_{ij} + a_{ij} \, B_j / C_{rcj}} \tag{4}$$

In effect, this assumption concentrates predation activity into smaller relative areas when C (foraging arena size) is small, so as to drive down vulnerable prey densities V_{ij} more rapidly as B_j increases in locales with less foraging arena area.

Importantly, including C_{rcj} as a modifier in the $a_{ij} \cdot B_j / C_{rcj}$ predation rate term results in the equilibrium predator biomass (B_j for which $dB_j/dt=0$) being proportional to C_{rcp}, i.e.,

$$B_j = \left(g_j \, v_{ij} \, B_i / Z_j - 2 \, v_{ij} / a_{ij} \right) C_{rcj} \tag{5}$$

That is, using the C_{rcj} as modifiers of the foraging arena consumption rate equation results in spatial patterns of biomass of consumers being proportional to C_{rcj}, other factors (prey biomasses B_i and mortality rates Z_j) being equal over space. We could, of course, also had assumed that variation in habitat capacity also affects the vulnerability exchange rates v_{ij}, search rates a_{ij}, and predation rates Z_j (and if so, added minor changes to the code to implement these assumptions), but the default assumption is that the dominant cause of "poor" or relatively small habitat capacity is lack of usable foraging arena area. As such, the basic change made to the rate equations is a simple division of the denominator terms for predator search term, by-arena vulnerable prey density equations, by the capacity values C_{rcj}.

The new model is made compatible with earlier Ecospace models by providing the option to derive capacity directly from presence/absence of habitats. In this case, habitat maps and habitat preferences are directly converted to a capacity map for each functional group. Cells that contain a preferred habitat will receive a full capacity of 1, other cells will receive a capacity of (almost) 0. The implementation in Ecospace further ensures that it is optional for every group in a model to use habitat maps and/or habitat capacity to drive distributions.

Setting initial adjusted biomasses

In going from Ecopath to Ecospace, it is assumed that the Ecopath base biomasses represent the average over all modeled cells of the cell-specific biomasses. This means that Ecospace biomass densities can be much higher in favourable spatial cells if there are relatively few such cells. Initial biomass densities $B_{rcj}(0)$ reflecting the C_{rck} variation are assigned at the start of each *Ecospace* simulation by assuming that these biomasses are proportional to the C_{rcj}. If there are nw water cells, such that overall biomass density for group k across the grid is given by nwB_i* where B_i* is the Ecopath base biomass for group j, the initial spatial biomass densities are assigned as

$$B_{rcj}(0) = (C_{rcj}/T\,C_j)\,nw\,B_j^* \tag{6}$$

where TC_j is a total capacity index over the grid for group j, i.e.,

$$TC_j = \sum_{r,c} C_{rcj} \tag{7}$$

and the sum over r and c is over all nw water cells in the spatial grid. Note that $TC_j \ll nw$ implies severe concentration of group j biomass on few cells.

Correction of search rate and vulnerability parameters for spatial overlap patterns

Spatial concentration of biomass for any group implies a requirement to adjust the rates of effective search au and vulnerability exchange rates vu for all foraging arenas u that are used by group j and its predators j', since without such adjustments predicted predation rates (using foraging arena equations from Ecosim) at the higher local densities would be artificially increased from the rates implied by Ecopath base consumptions. In order to make this adjustment, the rates are set so that the total consumption for each arena link is the same in Ecospace as in Ecosim, scaled up to the total number of water cells. This implies the condition

$$n_w \, Q_u = (a_u \, v_u \sum_{r,c} B_{rcj} \, B'_{rcj}) \, / \, (2 \, v_u + a_u \, B_j^*) \tag{8}$$

Here, Q_u is the Ecosim base biomass flow rate for arena link u, B'_{rcj} is the initial predator abundance (biomass for non-stanza groups or sum of numbers at age times length squared for multi-stanza groups) for cell (r,c), and B_{ju}^* is the spatially invariant initial predator abundance obtained by noting that applying the C_{rcj} correction in Eq. 4 results in

$$B_j^* = (n_j \, B'_{Ecosim,j})/TC'_j \tag{9}$$

Here $B'_{Ecosim,j}$ is the Ecosim initial predator abundance. Using the assumed relationships above, between initial B_{rcj}, B'_{rcj} and C_{rcj}, Eq. 8 can be written as:

$$Q_u = (a_u \, v_u \, B_{ju}^{**} \, B_j^*) \, / \, (2 \, v_u + a_u \, B_j^*) \tag{10}$$

where B_{ju}^{**} is the prey-predator "incidence weighted" mean prey biomass divided by B_j^* for link u given by

$$B_{ju}^{**} = (B_j^* \sum_{r,c} C_{rcj} \, C'_{rcj})/TC_j \tag{11}$$

Note that this reduces to just B_{ju}^* if all predator C'_{rcj} are near 1.0 for the same (r,c) cells where prey C_{rcj} are near 1.0, i.e. where there is strong spatial overlap of the prey and predator distributions, but can be much lower than for cases where predators occupy restricted spatial areas compared to the prey. Assuming the same vulnerability exchange rate vu as in Ecosim (from total base consumption rate over all predators using arena u) where k is the user-supplied vulnerability multiplier (aka *Vulmult*, see vulnerability chapter), Eq. 10 can be solved for a_u

$$a_u = (2 \, v_u) \, / \, [B_{j'}^* \, v_u \, B_{ju}^{**} \, / \, (Q_u - 1)] \tag{12}$$

Unfortunately, this calculation fails if $v_u \, B_{ju}^* \, / \, Q_u < 1$, which can happen with relatively low vu settings and weak overlap between prey and predator such that B_{ju}^* is much less than B_j. In that case, the assumed spatial distribution overlap pattern simply cannot support the total predation rate estimated for the link in Ecopath and Ecosim, and instead we simply set

$$v_u = 1.001 \, Q_u/B_{ju}^{**} \tag{13}$$

before solving for a_u in Eq. 12 so as to provide at least some large estimate of a_u to make simulations come as close as possible to predicting the base Q_u. The rate of effective search a_u is further adjusted upward by the multiplicative factor $Q_m \, Q_{oj'}/(Q_m \, Q_{oj'}-1)$ to account for handling time effects in order to create type II functional response effects by setting a low ratio of maximum (Q_m) to base feeding rate (Q_o).

Modification of spatial mixing rates to reflect movement toward preferred cells

For species with body sizes and mobility large enough to exhibit oriented dispersal and/or migration, it is reasonable to assume that dispersal rates between adjacent spatial cells are distorted so as to maintain abundance differences reflective of differences in habitat capacities between the cells.

Without such distortions or oriented movement, random dispersal between cells would greatly reduce abundance gradients created by the C_{rcj} capacity effects, and for species with restricted habitat use would result in too much biomass dispersing into unsuitable spatial cells so as to cause biomass to decrease substantially from Ecopath base biomasses, even without any changes in fishing pressure or predator abundances. For each border between cells, e.g. between cell (r,c) and cell $(r,c+1)$ to its right, *Ecospace* assumes instantaneous mixing rates $m_{1j}B_{rcj}$ to the right and $m_{2j}B_{rcj}$ to the left. Absent orientation implies $m_{1j}=m_{2j}=m_j$, where m_j is an (input) expected dispersal rate.[3] In order to avoid smearing of the distribution, the dispersal rates are set so that

$$m_{1j} B_{rcj} = m_{2j} B_{rc+1j} \tag{14}$$

Assuming biomasses are then to remain near or proportional to C_{rcj}, this balanced movement condition implies that the m_{1j} and m_{2j} have to be varied so as to meet the balance condition

$$m_{1j} / m_{2j} = C_{rc+1j} / C_{rcj} \tag{15}$$

Ecospace meets this condition by setting the exit rate to m_j for whichever cell has lower capacity C_{rcj}, then adjusting the exit rate for the cell with higher C_{rcj} to m_j times the capacity ratio. Thus for example if $C_{rc+1j} > C_{rcj}$, m_{1j} to the right is set to m_j and m_{2j} to the left is set to $m_j C_{rcj}/C_{rc+1j}$ so that m_{2j} will be very small if $C_{rcj} << C_{rc+1j}$, i.e. movement into the low capacity cell will be severely restricted.

Rounding off this chapter, the source publication[4] study used simulation modeling to evaluate the sampling characteristics of the habitat capacity model, based on an artificial data set and a spatial food web model of a marine ecosystem. This was used to derive "true" distribution based on environmental preference for the functional groups in the model, and then evaluate the degree to which it is possible to recreate the "true" distributions from sampling. As part of this, the impact of sample size and uncertainty in key parameters was evaluated. We refer to the source publication for details, and note that the habitat suitability model can be used to address a suite of new ecological questions, such as the impact of habitat degradation due to coastal development, eutrophication and climate change. In most cases, it should be considered to use the habitat capacity facility instead of the pre-defined habitat approach.

Attribution This chapter is based on Christensen, V, M Coll, J Steenbeek, J Buszowski, D Chagaris, and CJ Walters. 2014. Representing variable habitat quality in a spatial food web model. Ecosystems 17(8): 1397-1412. https://doi.org/10.1007/s10021-014-9803-3. Reused with License Number 5757230625588 from Springer Nature. Rather than citing this chapter, please cite the source.

Media Attributions

- From Figure 1 in Christensen et al. 2004

Notes

1. Christensen, V, M Coll, J Steenbeek, J Buszowski, D Chagaris, and CJ Walters. 2014. Representing variable habitat quality in a spatial food web model. Ecosystems 17(8): 1397-1412. https://doi.org/10.1007/s10021-014-9803-3

2. Steenbeek, J., Coll, M., Gurney, L., et al., 2013. Bridging the gap between ecosystem modeling tools and geographic information systems: Driving a food web model with external spatial–temporal data. Ecological Modelling 263, 139–151. https://doi.org/10.1016/j.ecolmodel.2013.04.027.

3. There is an IBM model dispersal rate estimator at https://ecopath.app

4. Christensen et al. 2014. *op. cit.*

47.

Spatial implementation of multi-stanza and IBM

Ecospace biomass dynamics

For functional groups not represented by multi-stanza population dynamics accounting, Ecospace represents biomass (B) dynamics over a set of spatial cells (k) with the spatially discretized rate formulation

$$dB_{ik}/dt = g_i Q_{ik} - Z_{ik} B_{ik} - \left(\sum_k m_{ikk'}\right) B_{ik} + \sum_{k'} m_{ik'k} B_{ik'} \qquad (1)$$

where B_{ik} is the biomass of functional group i in spatial cell k; g_i is conversion efficiency of food intake by group i into net production; Q_{ik} is total food consumption rate by group i in spatial cell k; Z_{ik} is total mortality rate of group i biomass due to predation, fishing, etc.; $m_{ikk'}$ is instantaneous movement rate of group i biomass from cell k to cell k'; and $m_{ik'k}$ is movement rate of group i biomass from cell k' to cell k.

All of the terms on the right hand side of Eq. 1, except g_i, are treated as dynamically variable over time so as to reflect changes in food availability (Q_{ik}), fishing effort and predation risk (Z_{ik}), and seasonal changes in movement patterns ($m_{ikk'}$). Food consumption rates Q_{ik} are calculated as sums over prey types j (i.e., $Q_{ik} = \Sigma j Q_{jik}$). Likewise, total mortality rates are calculated as sums over predator types and fishing fleets f: $Z_{ik} = M_{oi} + \Sigma_f F_{ifk} + \Sigma_j Q_{ijk}/B_{ik}$, where M_{oi} is unexplained mortality rate, the fishing rate components F_{ifk} by fleets f are predicted from spatial distributions of fishing effort for each "fleet" f over the grid cells k, and the Q_{ijk}/B_{ik} ratios represent predation rate components of M (i.e., $M_{ijk} = Q_{ijk}/B_{ik}$) calculated from predator j consumption rates Q_{ijk}.

The Ecospace grid cells are arranged as a rectangular grid with rows r and columns c, so that each cell k exchanges biomass directly only with those cells k' that are in adjacent rows and columns. If cell k represents row r, column c, then k' is restricted to cells ($r -$ 1,c), ($r + 1,c$), ($c - 1,r$), and ($c + 1,r$). Exchanges at the map perimeter are set to zero, except for groups that are assumed to be advected across the map, in which case biomasses at the map boundary are set to constant (Ecopath base estimate) values.

From the Ecosim Age-structured dynamics chapter:

The basic accounting relationships for multi-stanza groups are

$$N_{a+1,t+1} = N_{a,t} \exp(-Z_{s,t}/12) \tag{2}$$

$$W_{a+1,t+1} = \alpha_a q_{a,t} + \rho W_{a,t} \tag{3}$$

$$B_{s,t} = \sum_{a=a1(s)}^{a2(s)} N_{a,t} W_{a,t} \tag{4}$$

Here, $N_{a,t}$ is number of age a (in months) animals in calendar month t, $W_{a,t}$ is mean body weight of age a animals in month t, and $B_{s,t}$ is the biomass of stanza s, defined as the mass (numbers × weight) of animals aged $a1(s)$ through $a2(s)$ months. $Z_{s,t}$ is the total mortality rate of stanza s animals, defined the same way on the basis of fishing and consumption as for other model biomass groups i as $Z_{s,t} = M_{os} + \Sigma_f F_{sf} + \Sigma_j Q_{sj}/B_s$. All animals in stanza s are treated as having the same predation risk and vulnerability to fishing. The aggregated bioenergetics parameters a_a and r are calculated to make body growth follow a von Bertalanffy growth curve (with length-weight power 3.0) with user-defined metabolic parameter K. Exact von Bertalanffy growth occurs when predicted per-capita food intake $q_{a,t}$ is equal to a base food intake rate that is calculated from the consumption per biomass parameter (Q_s/B_s) provided by the user for each stanza. The metabolic parameter r, which equals $\exp(-3K/12)$, is based on the assumption that metabolism is proportional to body weight[1].

Actual or realized food intake $q_{s,t}$ at each time step is calculated from the total predicted food-intake rate for the stanza ($Q_{s,t}$) as $q_{s,t} = Q_{s,t} w_{a,t}^{2/3}/P_{s,t}$, where $P_{s,t}$ is the relative total area searched for food by stanza s animals and is computed as $P_{s,t} = \Sigma_a N_{a,t} w_{a,t}^{2/3}$. For foraging-arena food-intake and predation-rate calculations involving stanzas, $P_{s,t}$ is used instead of B_s as the predictor of total area or volume searched for food per unit time. The assumption that area searched and food intake vary as the ⅔ power of weight (i.e., as the square of body length) is a basic assumption that also underlies the derivation of the von Bertalanffy growth function. For notational simplicity, Eqs. 2-4 above are presented without a species index.

Original representation of multi-stanzas in Ecospace

When the multi-stanza option was originally developed for Ecosim, it was not incorporated directly into Ecospace. Instead, each stanza was treated as its own higher-order functional group for Ecospace biomass-dynamics calculations without accounting for age structure within the stanza. Rather, the age-structure of each stanza was assumed to be in equilibrium. Body weight was computed grid-wide (not cell-specifically) for each stanza. Feeding rates were assumed proportional to a relative search-area index P_s calculated from a prediction of the numerical abundance of the stanza N_s as $P_s = N_s \bar{P}_s$, where $\bar{P}_s$ is the initial (t = 0) per-capita mean of the relative area-searched index $P_{s,t}$, i.e.,

$$\bar{P}_s = \sum_a N_{a,0} W_{a,0} / \sum_a N_{a,0} \qquad (5)$$

Dynamics of the numbers in each stanza N_S were computed for each cell by the differential equation

$$dN_{sk}/dt = R_{sk} - Z_{sk}N_{sk} - \left(\sum_{k'} m_{ik'k}N_{sk}\right) \qquad (6)$$

where R_{sk} is an approximate difference between recruitment (incoming) rates and exit (to next stanza) rates for stanza s in spatial cell k. If the age structure within the stanza is assumed to remain near equilibrium, the R_{sk} term in Eq. 6 can be approximated as

$$R_{1k} = E_{tk}(1 - exp(Z_{sk}(a_2(1))/12)) \text{ for } s = 1 \qquad (7a)$$

$$R_{sk} = N_{s-1,k}Z_{s-1,k}/(\exp(Z_{s-1,k}(a_2(s-1) - a_1(s-1))/12) - 1) \text{ for } s > 1 \qquad (7b)$$

Eq. 7a represents egg production rate minus survival rate to the age at exit from stanza $s = 1$. Egg production is assumed to be approximately proportional to biomass B_{sk} of the oldest (adult) stanza s in cell k. Eq. 7b is derived from the equilibrium of the delay-differential equation for N_S that results from assuming spatial gain and loss rates to be approximately balanced, so that the dominant effects on N_S are gains from individuals progressing from the previous stanza and from losses of individuals as they progress to the next stanza and mortality within the stanza.

The equilibrium assumption needed for derivation of Eq. 7 can lead to inaccurate predictions because it can result in incorrect size distributions if incoming and outgoing numbers are not in balance, and size then affects the predation-rate parameters (areas searched, maximum prey-consumption rates). Eq. 7 is a relatively poor approximation for both egg production and net rates of numbers gained through graduation from younger stanzas and loss to older stanzas, so this early version of Ecospace tended to predict incorrect absolute values for cell-specific numbers N_{sk} relative to Ecospace-predicted cell-specific biomasses B_{sk}, but the predicted spatial distributions of abundances were at least qualitatively reasonable. In past applications, Ecospace generally predicted that N_{sk} was relatively high in cells with high egg production, in cells with favourable habitat, and in cells near seasonally varying optimum migration positions for migratory stanzas.

Predicting multi-stanza spatial distribution from continuous mixing-rate models

In an effort to avoid the large computer-memory requirements and massive accounting calculations (for typical models, on the order of 10^3 more calculations per time step) required for replicating the full age-structure accounting for multi-stanza groups for every grid cell of large Ecospace models, a simple approach based on combining the overall multi-stanza population accounting (as described in the Ecosim Age-structured dynamics chapter, eq. 2-4 above) with the relatively simple Eq. 7 diffusion model for predicting relative spatial abundances by stanza. This approach depends on two key assumptions:

1. the Ecosim multi-stanza accounting system can be applied for each multi-stanza population as a whole (summed over all Ecospace grid cells), given reasonable estimates of mean food consumption rates $q_{s,t}$ and mortality rates $Z_{s,t}$ averaged over the grid cells (a basic assumption that is made anyway in the non-spatial Ecosim representation of any large area) and

2. the diffusion model, Eqs. 6-7, gives reasonable predictions of the relative distribution of the biomass of each stanza over grid cells whether or not the absolute numbers N_{sk} are predicted correctly, hence preserving effects of complex spatial-overlap patterns among stanzas.

We then perform the Ecospace time solution on monthly time steps using the following four-step procedure:

1. We use the results from integration of Eqs. 6-7 to apportion the spatial distribution of total stanza biomass B_{st} over spatial cells k to give B_{sk} relative cell biomasses using $B_{sk} = B_{st}N_{sk}/\Sigma_k N_{sk}$.

2. The spatial B_{sk} biomasses (and relative predator-search areas $P_{sk} = P_{st}N_{sk}/\Sigma_k N_{sk}$) are then used in the Ecosim foraging arena and fishing rate calculations for each cell k to predict food-consumption rates Q_{sk} and mortality rates Z_{sk}.

3. Biomass-weighted average food-consumption rates $\bar{q}_{s,t}$, and mortality rates $\bar{Z}_{s,t}$, for the whole population are calculated as,

$$\bar{q}_{s,t} = \sum_k B_{sk}q_{sk}/B_{st} \text{ and } \bar{Z}_{s,t} = \sum_k B_{sk}Z_{sk}/B_{st} \qquad (4a,b)$$

4. The system-scale multi-stanza accounting is done by means of Eqs. 2-4 with the biomass-weighted averages $(\bar{q}_{s,t}, \bar{Z}_{s,t})$ to give predicted total population age and size structure and total stanza biomasses $B_{s,t+1}$ at the start of the next month.

This procedure retains some information about predicted changes in spatial abundance patterns due to mixing processes and spatial variation in mortality rates Z_{sk} because Z_{sk} is included in the prediction of relative numbers N_{sk} by cell from Eqs. 6-7, but it discards information about spatial variation in growth rates q_{so} in favour of using a single system-scale prediction of body growth (Eq. 3 with consumption rate $q_{s,t}$ represented by $\bar{q}_{s,t}$).

Further, it fails to account for the cumulative divergence that can take place in both age and size structure for relatively sedentary species resident in spatial cells that are protected from fishing. That is, for "adult" stanzas containing many age classes, it fails to represent the potential accumulation of older, more fecund animals in protected areas, considered by some to be a key benefit of marine protected areas. One possible solution to allowing accumulation of large adults in specific areas is to split the oldest stanza group into a number of stanzas, but doing so is an approximate fix rather than a solution.

For resident species, the mixing-model approach also fails to account for regional variation in growth rates associated with spatial cells that have higher basic (primary and lower-trophic-level) productivity or reduced intraspecific competition due to limited recruitment. For these reasons, the mixing-model approach is best suited to analyses of pelagic systems, where relatively high mobility results in averaging of feeding and mortality rates over substantial areas. When used for systems with many resident or sedentary species, the approach is potentially misleading and should be used only to provide computationally "quick-and-dirty" policy screening for options such as size and spacing of MPAs, to be followed by more careful screening according to the more detailed individual-based approach described below.

Individual-based approach for predicting spatial patterns in growth, survival, and distribution

Most regional populations exhibit at least some degree of localized or cell-scale variation in recruitment, body growth, and survival rates, and erosion of this local structure has serious implications for maintenance of both biodiversity and overall productivity. The original and mixing-model approaches described above cannot adequately capture such local structure, which can result from the cumulative effects of the development of a fishery or from MPAs. Ecospace therefore includes a much more detailed and realistic approach to the representation of localized trophic-interaction effects based on concepts of individual-based modeling (IBM).

In the IBM approach, we retain the spatial biomass-dynamics accounting for non-multistanza species represented by Eq. 1 and the multi-stanza population dynamics accounting of Eqs. 2–4, but rather than solving Eqs. 2–4 once for each stanza using spatially averaged (grid-wide) food-consumption and mortality rates, we divide the age-0 recruits for each multi-stanza population ($N_{0,t}$) into a large number np of packets (cohorts). Each packet is assumed to represent some number of identical individuals of the population, and all packets from the monthly recruitments start out with the same individual biomass and numbers at recruitment ($N_{p,0,t} = N_{0,t}/n_p$). Each packet is then followed independently as it moves among spatial cells on the grid. This approach is similar to that recommended by Rose et al.[2] and Scheffer et al.[3]

The growth-survival Eqs. 2-3 are then solved for each packet, yielding its predicted age and size dynamics ($N_{p,a,t}$ and $W_{p,a,t}$). Packets are discarded from the overall population when they reach a maximum age (denoted amax) beyond which $N_{p,a,t}$ is negligible. Each packet p is assigned an initial spatial position $X_{p,0,t}, Y_{p,0,t}$, and movements of the packet over time are predicted from both random (diffusive) and oriented (migratory) changes in position. At each simulation time step, the ecological conditions (food intake rates, mortality rates) for the spatial cell in which each packet is located are used in Eqs. 2–3. The overall accounting for $B_{s,k}$ and P_{sk} needed for trophic-interaction predictions (impacts from and on biomasses of non-stanza species in each cell k) then involves simply summing $B_{p,k,t}$ and $P_{p,k,t}$ packet biomasses and predation search areas over those packets present in each cell k, before foraging arena predictions of Q_{sk}, Z_{sk} are performed for that cell.

The obvious advantage of the IBM approach is that it retains the cumulative history of each packet's space-use pattern, in the form of the packet's numerical (worth) and body size (weight) states. For sedentary species, local differentiation in growth and accumulation of older animals is represented by how packets in different local areas (cells) fare over time. Further, through use of restricted movement rules, collections of packets can easily be made to form distinctive local populations, presumably key units of local adaptation and biodiversity, including as a consequence of local environmental conditions.

A disadvantage of the approach is that it requires massive computation, both for the survival-growth calculations and for movement of a sufficient number of packets over the simulated grid to permit realistic spatial distributions and variation. This number must be determined by trial and error; the number of packets must be increased until results stop changing. Most of the computational effort (typically about 90%) ends up being in the simulation of movement as changes in the locations $X_{p,a,t}$, $Y_{p,a,t}$.

Monthly survival-bioenergetics updates for each packet are based on food intake and mortality rates predicted for the spatial cell where the packet is located at the start of the month. No attempt is made to integrate q or Z rates over times within the month spent in different cells; doing so would be prohibitively computationally intensive. This omission amounts to assuming either that cell sizes are set large enough that most movements over any month occur within a single cell or that spatial correlation in productivity and predation risk among nearby cells there is reasonably high, so movements over such cells would result in the same predicted food intake and mortality rates obtained from the initial cell. Effects of violating this assumption could be tested if the model were run with varying grid cell sizes.

The initial or spawning position for each packet ($X_{p,0,t}$, $Y_{p,0,t}$) is set to the centre of a cell k, where the probability of recruiting to cell k is set equal to $E_{kt}/\Sigma_k E_{kt}$ and E_{kt} is the predicted total egg production in cell k for month t summed over all packets that are in cell k at the start of the month. This procedure allows spawning to occur well away from locations of larval settlement or juvenile growth because larval dispersal and juvenile migration can be explicitly represented, through either different or similar movement-simulation rules as used for packets of older fish. In particular, the IBM approach "encourages" formation of local stock structure; recruitment tends to occur near centres of egg production. In the context of MPAs, lower mortality rates Z in designated cells can result in the accumulation of older, more fecund fish, and those cells can thus become local areas of high reproduction.

Monthly movements by each packet are simulated as a set of ns increments to the X,Y values that determine location on the grid. The user specifies an average annual movement distance, which implies an average monthly movement distance. The number of moves ns each month is then set so that the distance per increment cannot exceed the width of one cell. Each movement is made only in a cardinal direction (N,S,E,W), so that only X or Y (not both) changes for each move. The probability of choosing each of the four directions, k', is set to $m_{skk'}/\Sigma_{k'} m_{skk'}$, where $m_{skk'}$ is the instantaneous movement rate from cell k to k' calculated for the continuous biomass model (see Eq. 1). As noted above, the $m_{skk'}$

can be set equal for all k', to represent purely diffusive movement, or biased to represent avoidance of cells with unsuitable habitat, movement toward preferred habitats, or seasonal migration patterns. This method for choosing movement directions makes it possible to use the same user interface for entering assumptions about movement distances and orientation for multi-stanza populations as for groups represented only by biomasses, and it ensures that the multi-stanza movement patterns are broadly comparable with predictions from the computationally faster continuous mixing-model version of Ecospace.

Attribution The chapter is adapted from Walters et al. 2010, *Bulletin of Marine Science*[4], which permits authors to use figures, tables, and brief excerpts in scientific and educational works provided that the source is acknowledged and the use is non-commercial.

Notes

1. Essington, T. E., J. F. Kitchell, and C. J. Walters. 2001. The von Bertalanffy growth function, bioenergetics, and the consumption rates of fish. Can. J. Fish. Aquat. Sci. 28: 2129–2138. https://doi.org/10.1139/f01-151

2. Rose, K. A., S. W. Christensen, and D. L. DeAngelis. 1993. Individual-based modeling of populations with high mortality: a new method based on following a fixed number of model individuals. Ecol. Model. 68: 273–292. https://doi.org/10.1016/0304-3800(93)90022-K

3. Scheffer, M., J. M. Baveco, D. L. DeAngelis, K. A. Rose, and E. H. Van Nes. 1995. Super-individuals: a simple solution for modeling large populations on an individual basis. Ecol. Model. 80: 161–170. https://doi.org/10.1016/0304-3800(94)00055-M

4. Walters, C., Christensen V, Walters W, Rose K. 2010. Representation of multi-stanza life histories in Ecospace models for spatial organization of ecosystem trophic interaction patterns. Bull. Mar. Sci. 86(2):439-459

48.

Spatial fishery dynamics

Fishing fleets are specified in the Ecopath model, along with landings, discards, discard mortality rates, discard fate and market values of each landed species and non-market price. When moving to Ecospace, it is important to enter the percentage of costs that are "sailing related" (e.g., fuel, crew) in Ecopath and then specify how the relative cost of fishing is expressed across the modeled area. With this information, Ecospace will distribute fishing effort spatially through a gravity model[1][2], where the effort allocated to each spatial cell is based on the profitability of fishing estimated as the difference between expected income and costs of fishing in each cell[3], thus proportional to the net benefits (profits-costs) gained from fishing in a given cell. When no cost or revenue information is entered, the fleets will gravitate to cells with the highest biomass of their target species. The cost of fishing can also be used to keep fleets from operating in certain areas, e.g., on the windward side of Caribbean islands or to keep fleets out of neighboring countries' EEZ.

Ideally, expected income should be estimated with local economic and market data, otherwise it can be estimated using prices per functional groups from global price databases available based on average prices[4], or even better, regional or local data if applications are regional.

There is an option on the Ecospace map to calculate the cost of fishing based on the assumption that it is proportional to the distance (km) from the nearest coast or port. This is done based on Euclidean distances from the nearest port for each fleet. The calculation is very rudimentary and does not calculate pathway lengths around land masses thus, more suitable data can be used when available.

Each fleet can be allowed to operate – or restricted from fishing – in one or more habitat types, and MPAs can be placed that exclude one or more fishing fleets during specified months. For a detailed description of fishing dynamics in Ecospace,[56]. In addition, there is a new feature of the Ecospace model called the "MPA dynamic routine". This routine allows one to study the spatial-temporal creations of MPAs, which can be allocated to fleet(s) at any moment in time during a simulation, and can be operative monthly or annually. This feature allows full control of MPA location or deletion in time and space[7]. It can be a good idea to use environmental preference functions to distribute ecosystem groups spatially and habitats to distribute fishing effort.

Ideally, the emergent fishing effort distributions predicted by Ecospace should be compared to observed spatial effort patterns, such as those from logbook and vessel monitoring systems (VMS and/or AIS) data[8], and parameters would be adjusted to improve agreement. However, this has rarely been done in practice[9], due to a lack of spatial effort data in most modeled systems and difficulties in modeling fisher behavior and complex spatial management. Alternatively, one may attempt to force fishing effort using the spatial-temporal framework by either directly importing fleet-specific sailing cost maps that make it prohibitively expensive to fish in certain cells, or by designating an MPA or habitat for each fleet and loading time-varying maps of closed cells or habitat types. Additionally, habitat layers and MPAs can be used to impose administrative boundaries that prohibit fleets from operating in certain jurisdictions. While Ecospace contains great flexibility in representing spatial fishery dynamics, thus far little attention has been devoted to spatial effort dynamics[10,11] compared to the biomass and ecological responses generated by Ecospace. This represents a component of the Ecospace that would benefit from simulation research, demonstration, and application.

Attribution

The chapter is based on de Mutsert et al.[12], adapted with permission.

Notes

1. Caddy, J.F. 1975. Spatial model for an exploited shellfish population, and its application to the Georges Bank scallop fishery. J. Fish. Res. Board Can. 32: 1305–1328. https://doi.org/10.1139/f75-15

2. Walters, C.J. and R. Bonfil, 1999. Multispecies spatial assessment models for the British Columbria groundfish trawl fishery. Can. J. Fish. Aquat. Sci. 56:601- 628. https://doi.org/10.1139/f98-205

3. Walters, C., Pauly, D., Christensen, V., 1999. Ecospace: Prediction of Mesoscale Spatial Patterns in Trophic Relationships of Exploited Ecosystems, with Emphasis on the Impacts of Marine Protected Areas. Ecosystems 2, 539–554. https://doi.org/10.1007/s100219900101

4. Sumaila, U.R., Marsden, A.D., Watson, R., Pauly, D., 2007. A Global Ex-vessel Fish Price Database: Construction and Applications. J Bioecon 9, 39–51. https://doi.org/10.1007/s10818-007-9015-4

5. Walters, C., Pauly, D., Christensen, V., 1999. Ecospace: Prediction of Mesoscale Spatial Patterns in Trophic Relationships of Exploited Ecosystems, with Emphasis on the Impacts of Marine Protected Areas. Ecosystems 2, 539–554. https://doi.org/10.1007/s100219900101

6. Walters, C., Christensen, V., Walters, W., Rose, K., 2010. Representation of multistanza life histories in Ecospace models for spatial organization of ecosystem trophic interaction patterns. Bulletin of Marine Science 86, 439–459.

7. e.g., Gomei, M., Steenbeek, J., Coll, M., Claudet, J., 2021. 30 by 30: Scenarios to recover biodiversity and rebuild fish stocks in the Mediterranean.

8. Piroddi, C., Coll, M., Macias, D., Steenbeek, J., Garcia-Gorriz, E., Mannini, A., Vilas, D., Christensen, V., 2022. Modelling the Mediterranean Sea ecosystem at high spatial resolution to inform the ecosystem-based management in the region. Sci Rep 12, 19680. https://doi.org/10.1038/s41598-022-18017-x

9. see Romagnoni, G., Mackinson, S., Hong, J., Eikeset, A.M., 2015. The Ecospace model applied to the North Sea: Evaluating spatial predictions with fish biomass and fishing effort data. Ecological Modelling 300, 50–60. https://doi.org/10.1016/j.ecolmodel.2014.12.016

10. Martell, S.J.D., Essington, T.E., Lessard, B., Kitchell, J.F., Walters, C.J., Boggs, C.H., 2005. Interactions of productivity, predation risk, and fishing effort in the efficacy of marine protected areas for the central Pacific. Can. J. Fish. Aquat. Sci. 62, 1320–1336. https://doi.org/10.1139/f05-114

11. Adebola, T., De Mutsert, K., 2019. Spatial simulation of redistribution of fishing effort in Nigerian coastal waters using Ecospace. Ecosphere 10, e02623. https://doi.org/10.1002/ecs2.2623

12. De Mutsert K, Marta Coll, Jeroen Steenbeek, Cameron Ainsworth, Joe Buszowski, David Chagaris, Villy Christensen, Sheila J.J. Heymans, Kristy A. Lewis, Simone Libralato, Greig Oldford, Chiara Piroddi, Giovanni Romagnoni, Natalia Serpetti, Michael Spence, Carl Walters. 2023. Advances in spatial-temporal coastal and marine ecosystem modeling using Ecopath with Ecosim and Ecospace. Treatise on Estuarine and Coastal Science, 2nd Edition. Elsevier. https://doi.org/10.1016/B978-0-323-90798-9.00035-4

49.

Predicting spatial effort

EwE works with multiple fishing fleets, with fishing mortality rates (F) initially distributed between fleets based on the distribution in the underlying Ecopath base model. In Ecospace the F's are distributed using a simple logit-choice or "gravity model" where the proportion of the total effort allocated to each cell is assumed proportional to the sum over groups of the product of the biomass, the catchability, and the profitability of fishing the target groups, divided by relative cost of fishing the cell [1][2]. This profitability of fishing includes factors such as the cell-specific cost of fishing.

Assuming that there are N cells representing water areas, each fleet k can cause a total fishing mortality rate $N \cdot F_k$. For each step in the simulation this rate is distributed among cells, c, in proportion to the relative utility weights G_{kc} calculated as

$$G_{kc} = O_{kc} U_{kc} \frac{\sum_i p_{ki} q_{ki} B_{ic}}{C_{kc}} \tag{1}$$

where O_{kc} is 1 if cell c is open to fishing by fleet k, and 0 if not; U_{kc} is 1 if the user has allowed fleet k to work in the habitat type to which cell c belongs, and 0 if not; p_{ki} is the relative price fleet k receives for group i fish, q_{ki} is the catchability of group i by fleet k (equal to the F_{ki} in the Ecopath model); B_{ic} is the biomass of group i in cell c; and C_{kc} is the cost for fleet k to operate in cell c. Based on the weights in Eq. 1 the total mortality rate is distributed over cells according to

$$F_{kc} = N\,F_k\,G_{kc}^p / \sum_c G_{kc}^p \tag{2}$$

while each group in the cell is subject to the total fishing mortality

$$F_{ic} = \sum_k F_{kc}\,q_{ki} \tag{3}$$

The p parameter here represents variation among fishers in perception of the best place to fish, and is set to 1.0 by default. Setting p to higher values results in effort being more concentrated in the most profitable cells, and lower values cause effort to be more spread out (due to either wide variation among fishers in their actual best locations to fish, or lack of information that causes them to just try fishing everywhere). Readers familiar with logit choice theory may recognize the G weights as exp(utility) values, with utility assumed to be proportional to the logarithmic difference ln(income)-ln(cost) in income and cost components of decision choices.

Attribution This chapter is in part adapted from the unpublished EwE User Guide: Christensen V, C Walters, D Pauly, R Forrest. 2008. Ecopath with Ecosim User Guide.

Notes

1. Caddy, J.F. 1975. Spatial model for an exploited shellfish population, and its application to the Georges Bank scallop fishery. J. Fish. Res. Board Can. 32: 1305–1328. https://doi.org/10.1139/f75-15

2. Hilborn, R., and Walters, C. J. 1987. A general model for simulation of stock and fleet dynamics in spatially heterogeneous fisheries. Canadian Journal Of Fisheries And Aquatic Sciences, 44(7):1366-1369 https://doi.org/10.1139/f87-163

50.

Migration and advection

Ecospace dynamically allocates biomass across a grid map while accounting for mixing rates between adjacent grid cells. In the simplest cases, the basic assumptions about these rates are,

1. they are symmetrical from a cell to its four adjacent cells if all four cells have equal habitat capacity,

2. which is modified whether a cell is defined as "preferred habitat" or not (running means over adjacent sets of five cells allows for smooth transitions between habitat types), and

3. user-defined increased predation risk and reduced feeding rate in non-preferred habitat.

Additionally, Ecospace can simulate advection of biomass for organisms that drift passively with surface currents, and also seasonal migrations of organisms that move over large distances within each year. For multi-stanza groups, an additional option (see Ecospace IBM chapter) allows division of each monthly cohort into a large number of packets, with movement of these packets simulated as a stochastic process so as to potentially generate realistic patterns of movement (and possibly migration) of organisms of different ages.

Representing seasonal migration in Ecospace

Larger organisms commonly have seasonal migration patterns that allow them to utilize favourable seasonal resource and environmental conditions over large spatial areas. Such movements can be represented in Ecospace in two ways. First is using a simple "Eulerian" approach, which involves explicitly modelling changes in instantaneous rates of biomass mixing among the Ecospace spatial cells, in some way that approximates at least the changing center of distribution of the migratory species. The second way is the "Lagrangian" approach, which is for multi-stanza groups only. It simulates stochastic movements of a large number of "individuals", i.e packets of biomass that move together over the spatial map.

The Eulerian approach is implemented in Ecospace by allowing users to define a monthly sequence of "preferred" position map cells (or clusters of cells) by first declaring which groups that are migratory on the *Ecospace > Input > Dispersal* form, then on the *Eco-*

space > Input > Maps form sketch (or import) for each month the preferred cells for each migration group.

The Migration dialogue box displays a map of the Ecospace region, with migratory species, month by month over a calendar year. Preferred position for each month (and the annual trajectory of preferred positions) is set by setting a value on the interface and sketching the area of distribution

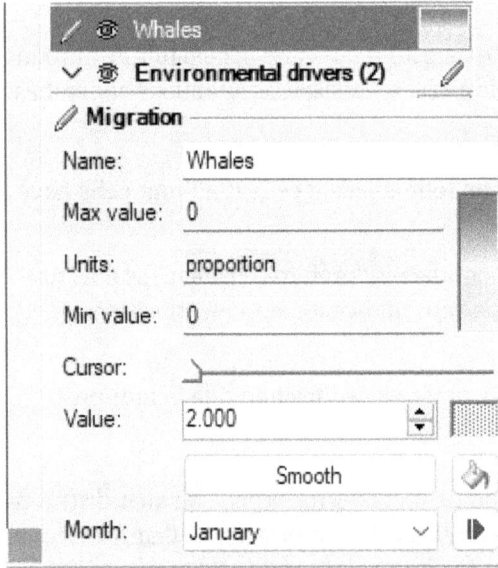

Figure 1. Migration input at *Ecospace* > *Input* > *Maps*.

Double-clicking on the selected functional group name ("Whales" in Figure 1), will bring up a spreadsheet with the migration values. These can be imported and exported, and thus derived externally for more reproducible behaviour.

The mathematical method used in Ecospace to create migratory behaviour is quite simple. Spatial movement is represented in general in Ecospace as a set of instantaneous exchange rates across the boundaries of adjacent spatial cells. For migratory species, these exchange rates are simply multiplied by relative factors at each simulation time step, where the factors depend on distance from the preferred cell for that time step as shown in Figure 2. The function is reversed for movement across a northern cell boundary. A similar function is used for east-west movements with map column-preferred column as the independent variable.

The factor has no effect (multiplies movement rates by 1.0, so movement rates are similar in all directions) for cells near the preferred cell (or cluster of cells), and "shuts down" movement away from the preferred cell for cells far from that preferred cell. Note that the base movement rates that are multiplied by the migration factors may not be the same in all directions to start with; these base rates can include advection effects and/or increased/ oriented movement rates towards preferred habitats. That is, migration effects can be combined with advection and orientation of movement toward preferred habitats; it was the intention to represent such combined effects that motivated the multiplicative factor formulation in the first place.

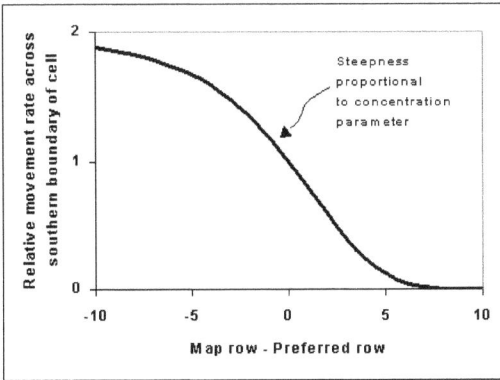

Figure 2. Relative movement rates; see text for details.

Tips for setting up migration in Ecospace

Unfortunately, there is no way to make the Ecospace migration simulations very simple to set up, particularly when using the IBM option to simulate movement more realistically for multi-stanza groups. Generally one must do considerable numerical experimentation to find reasonable migration parameter values and a stable numerical solution scheme. These cannot be computed in advance since they depend on a variety of details about the spatial map grid and species movement characteristics. Here are a few key points to keep in mind while experimenting (by repeated simulations) with the migration interface:

- The concentration parameters are relative values that the user needs to set by trying alternatives (generally in the range 0.5 to 4.0) to see what values give general distribution patterns similar to those observed in the field. Low values (<1.0) lead to weak distortion of movement toward preferred cells and hence to more widely spread distributions, while high values (e.g. 3.0) give distributions strongly concentrated near the preferred cells.

- Mean annual movement distances (*Ecospace > Input > Dispersal* form) have to be set large enough for migrating species to be able to "track" movements in preferred locations. As a general rule, set the base dispersal rate for migratory species to at least 100L km/year, where L is mean body length in cm. This is particularly critical for multi-stanza groups where different stanzas commonly reside in non-overlapping areas, and may need to move considerable distances during each "ontogenetic habitat shift" (without suffering too high relative predation risk or poor feeding rate conditions).

- Setting high concentration parameter values (>2.0) and/or moving animals through a very complex map with many coastal blocking features can result in numerical instability in the Ecospace solution algorithm. The best way to correct this is to reduce the movement distances somewhat; it may also be necessary to reduce the Successive over-relaxation (SOR) weight (*Ecospace > Input > Ecospace parameters* form) used in solving the linear equations involved in the numerical scheme for integrating the spatial rate equations (implicit method, BDF2 backward differentiation that is most often stable but can be problematic when there are very strong spatial gradients).

- Setting high concentration parameter values can also result in "overfishing". Ecospace allocates total fishing effort over the map proportional to the total number of cells initially used by each fishing fleet, so when the model generates a concentrated distribution of some favoured species, the total effort will concentrate accordingly and can sometimes generate very high fishing rates near the center of the migrating stock distribution. Remedies include reducing total effort by reducing the total efficiency multiplier *(Ecospace > Input > Ecospace fishery > Fleet dynamics > Tot. eff. multiple.)* and distributing effort more widely (reduce value of "effective power" on the same form).

- Concentrating a migratory predator can cause local depletion of food organisms and/or reduced per- predator feeding rates due to prey vulnerability limits. If these effects cause simulated total predator biomass to incorrectly decline over time (and if the user determines that the declines are not due to an artifactual overfishing effect), then it may be necessary to either increase total prey abundances (in Ecopath) or vulnerability of prey to the predator *(Ecosim > Input > Vulnerability multipliers* form).

- Multi-stanza population dynamics may behave strangely or incorrectly when one or more life history stages are migratory while other(s) are not. Ecospace does not keep track of the full population age/size distribution for each spatial cell (prohibitive memory and computing time requirement), and instead updates only the total abundances by stanza then distributes those using either the non-stanza prediction of biomass distribution or the IBM packet simulation approach. Either approximation tends to "dampen" abundance fluctuations in the early life history stanzas that might be created by, for example, seasonal movement of the adults to spawning locations near preferred juvenile habitats.

We were cautious above when describing the Ecospace potential to track migrating stanza populations, but a recent PhD dissertation by Fanny Couture (soon available at the UBC Open Collection of Theses and Dissertations), shows that it can be possible to successfully track migrating stanza of numerous different seasonal runs of Pacific salmon. The outcome of this was indeed beyond our cautious expectations.

Advection in Ecospace

Advection processes are critical for productivity in most ocean areas. Currents deliver planktonic production to reef areas at much higher rates than would be predicted from simple turbulent mixing processes. Upwelling associated with movement of water away from coastlines delivers nutrients to surface waters, but the movement of nutrient rich water away from upwelling locations means that production and biomass may be highest well away from the actual upwelling locations. Convergence (down-welling) zones represent places where planktonic production from surrounding areas is concentrated, creating special opportunities for production of higher trophic levels.

Once an advection pattern has been defined, the user can specify which biomass pools are subject to the advection velocities (vu,vv field) in addition to movement caused by swimming and/or turbulent mixing. This allows examination of whether some apparent "migration" and concentration patterns of actively swimming organisms, (e.g., tuna aggregations at convergence zones) might in fact be due mainly to random swimming combined with advective drift.

Older versions of Ecospace had an interface that allowed users to "sketch" simple advection patterns which were then corrected to insure mass balance (or allowed to exhibit areas of upwelling and downwelling into depths not modeled by Ecospace). That method never worked well, and we advise users to work with advection fields calculated with credible and well-tested hydrodynamic models.

Output from hydrodynamic models can be used as time-varying spatial input for Ecospace via the Temporal Spatial Framework of EwE, see the EwE User Guide for details.

Media Attributions

- Ecospace > Input > Maps > Migration

VIII

Spatial Planning

51.

Spatial planning: MSP Challenge

Jeroen Steenbeek

Marine or Maritime Spatial Planning (MSP) has developed as a tool to evaluate how the many and often conflicting uses of the marine environment may best be designed. However, management of marine space, at various scales, across a wide range of socioeconomic sectors and national and transnational boundaries, under the influence of natural processes and climate change, with their associated uncertainty, is a daunting task for which managers and planners are often ill-prepared[1].

As a reaction to this challenge, the Marine Spatial Planning (MSP) Challenge simulation platform was developed to integrate best available geographic, maritime, and marine data provided by many proprietary institutions with science-based simulation models for shipping, energy, and ecology. The data and models are linked together in a Unity game-engine based interactive platform[2]. This simulation platform allows anyone, experts as well as nonexperts, to operate it for planning support such as stakeholder engagement, codesign, interactive scenario development, professional learning, and student education.

The MSP Challenge simulation platform has been used for many interactive sessions with planners, stakeholders, and students in various parts of the world. In an interactive session or "game", participants take up the role of planner (or stakeholder) in one of the countries in a sea basin. As planners, users have an overview of the entire sea region and can review many different data layers to make an assessment of the current status. They design management plans for future uses of space in their exclusive economic zone over a period of several decades, and need to negotiate with neighbouring planners to have their plans approved. They can also consult other countries or develop and implement transboundary plans regarding a wide range of planning activities, including shipping routes, wind farms and power grids, and marine protected areas (MPAs). In a typical MSP session, planning phases alternate with simulation phases where the consequences of planning decisions for energy, shipping, and the marine environment are simulated and visualized as indicators and heat maps at the sea basin level. Digital game technology makes it fun and easy to draw and modify plans, run the simulations, and interact with others. Elements of gameplay such as challenges and objectives, a story line, role-play, and performance feedback can be used to facilitate interactions among the participants in a session.

The MSP Challenge simulation platform and the EwE approach have now been integrated to translate the gradual implementation of spatial plans into changes in environmental conditions and fisheries regulations (henceforth called pressures). These pressures were incorporated into the calculations of Ecospace to affect the state of the marine ecosystem components over time and space. Aggregated, spatially explicit Ecospace predictions (henceforth called outcomes) were sent back to the MSP Challenge to disseminate the state of the ecosystem components to session participants.

Both the MSP Challenge and the EwE approach are data-driven software systems. Within the bounds of a fixed set of equations and behaviours embedded within the software, these systems are parameterized to represent any ecosystem with its specific challenges and dynamics. We have connected the MSP Challenge and EwE software systems through a few new software components as shown in Figure 1.

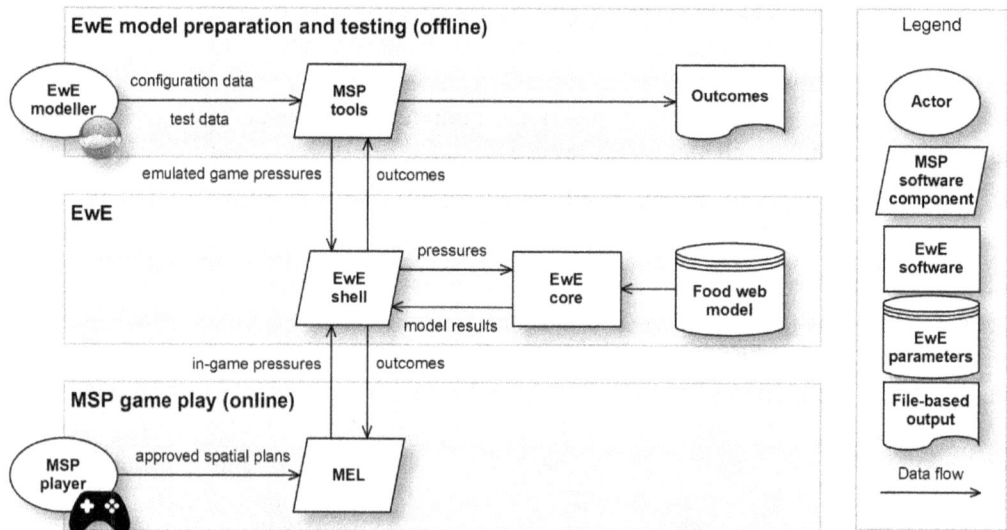

Figure 1. An overview of software components needed to connect the Ecopath with Ecosim (EwE) computational core to the marine spatial planning (MSP) challenge. EwE shell is the central software library that encapsulates the EwE modeling logic to receive MSP player-derived pressure layers and to deliver outcome layers back to the MSP software system. MSP Tools is a plug-in to the EwE desktop software that allows EwE modellers to design the connectivity between a specific MSP scenario and an EwE model, and to test the behaviour of this EwE model as if connected to the actual MSP game. MEL, the MSP-EwE Linker, is a software library that integrates EwE shell into the MSP game engine by converting MSP player actions to pressure maps for consumption by the EwE shell, and by delivering ecological outcomes to the MSP Challenge game.

Pressures: impacting the ecological model

Several distinct pressure categories were defined,

- **Noise** is the spatial distribution and intensity of low frequency noise resulting from shipping, construction, etc. The noise map layer acts as an environmental driver layer in the Ecospace habitat capacity model and affects per-cell foraging suitability for functional groups sensitive to low-frequency noise.

- **Surface disturbance** and **bottom disturbance** are the spatial distribution and intensity of physical disturbance at the surface and the bottom, respectively. This pressure includes the presence of temporary and transient structures and vehicles, turbidity due to anthropogenic activity, some forms of pollution, etc. The disturbance map layers act as environmental driver layers in Ecospace and affect per-cell foraging suitability for functional groups sensitive to these disturbances.

- **Artificial substrate** is the spatial distribution and intensity of artificial structures that provide shelter and/or habitat to sensitive functional groups. This layer acts as an additional habitat in Ecospace to increase habitat-derived cell suitability in Ecospace.

- **Protection** is the spatial distribution of locations where fishing is impossible due to the presence of other activities or prohibited through fisheries restrictions. This per-fleet map layer acts as a MPA layer in Ecospace, blocking fishing effort for all sensitive fishing gears in cells where MSP activities that generate this pressure are present.

- **Fishing intensity** is a scalar pressure to increase or decrease the nominal amount of fishing across the game area.

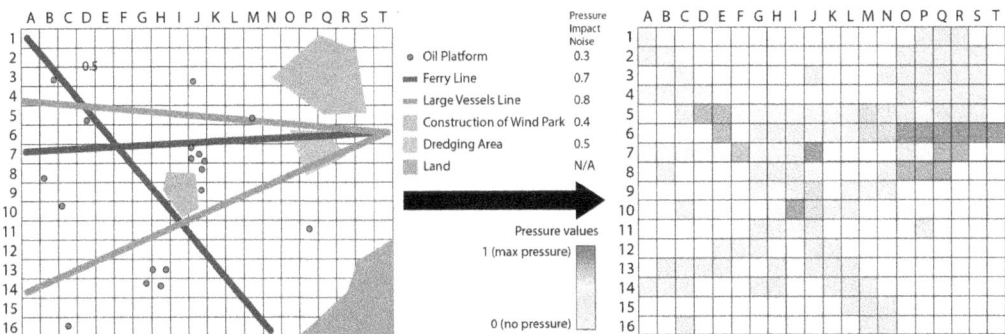

Figure 2. Schematic overview showing how spatial plans (actions), in vector format, are converted to pressure grids, in raster format. This example shows how oil platforms, ferry and shipping lines, wind park construction, and dredging contribute to the noise pressure grid via conversion factors unique to each type of action.

During simulation phases, the MSP Challenge software converts spatial plans created by players into pressure maps using an action-pressure conversion matrix, as follows (Figure 2).

- Spatial plans, which are entered as points (e.g., anchorages, oil and gas platforms) and lines (e.g., shipping routes, cables, pipelines, etc.), are spatially expanded to their area of impact using impact factors, expressed in the action-pressure conversion matrix as a ratio of Ecospace cell size. The zone width may be multiplied by the intensity of the spatial plan where applicable (e.g., shipping intensity). The zone-to-cell surface overlap is then calculated as a measure of spatially explicit pressure intensity on a value range from zero (no pressure) to one (maximum pressure).

- The cell area overlap of spatial plans, which are entered as polygons (e.g., dredging sites, marine protected areas, harbours, wind farms, etc.), is directly calculated and multiplied by the impact amount stated in the action pressure matrix and intensity of the spatial plan, where applicable.

- Total pressures from point, line, and polygon features are added per cell and range from zero (no pressure) to one (maximum pressure). Ecospace directly integrates the pressure grids into designated maps of environmental drivers, habitats, and protection, and directly incorporates the per-fleet fishing effort multiplier, to affect ecosystem dynamics.

Ecospace directly integrates the pressure grids into designated maps of environmental drivers, habitats, and protection, and directly incorporates the per-fleet fishing effort multiplier, to affect ecosystem dynamics.

Outcomes

Outcomes are spatially explicit aggregations of Ecospace predictions. The complex results of food web dynamics are condensed to provide MSP session participants with key ecological results and indicators. Outcomes can consist of four types of Ecospace predictions: group biomass, group catch, fleet effort, and biodiversity indicators.

The data in the outcome maps are reflected in the MSP software on a fixed colour gradient that represents one order of magnitude deviation from Ecopath baseline values. This relatively simple display system facilitates game participants to perceive drastic (local) changes in ecosystem functioning on a uniform scale across all outcomes.

Attribution The chapter is an extract from Steenbeek et al. 2020[3] adapted under a Creative Commons Attribution-NonCommercial 4.0 International License. Rather than citing this chapter, please cite the source.

Media Attributions

- Figure-2_EwE-desktop-tools
- Figure-3_Vector-to-raster-conversion_v2

Notes

1. Mayer, I., Zhou, Q., Lo, J., Abspoel, L., Keijser, X., Olsen, E., Nixon, E., Kannen, A., 2013. Integrated, ecosystem-based Marine Spatial Planning: Design and results of a game-based, quasi-experiment. Ocean & Coastal Management 82, 7–26. https://doi.org/10.1016/j.ocecoaman.2013.04.006

2. Abspoel, L., I. Mayer, X. Keijser, H. Warmelink, R. Fairgrieve, M. Ripken, A. Abramic, A. Kannen, R. Cormier, and S. Kidd. 2021. Communicating maritime spatial planning: the MSP Challenge approach. Marine Policy 132. https://doi. org/10.1016/j.marpol.2019.02.057

3. Steenbeek, J., G. Romagnoni, J. W. Bentley, J. J. Heymans, N. Serpetti, M. Gonçalves, C. Santos, H. Warmelink, I. Mayer, X. Keijser, R. Fairgrieve, and L. Abspoel. 2020. Combining ecosystem modeling with serious gaming in support of transboundary maritime spatial planning. Ecology and Society 25(2):21. https://doi.org/10.5751/ES-11580-250221

52.

Spatial optimization

This chapter provides a brief introduction to the Spatial optimizations tool in Ecospace, (*Ecospace > Tools > Spatial optimizations*). For instructions on how the routine is implemented in EwE, see Spatial optimizations in the EwE User Guide. There is also a tutorial in the next chapter (web- and pdf-versions).

We describe two approaches for spatial optimization of protected area placement, both based on maximizing an objective function that incorporates ecological, social, and economic criteria. Of these, a seed cell selection procedure[1] works by evaluating potential cells for protection one by one, picking the one that maximizes the objective function, add seed cells, and continue to full protection. The other is a Monte Carlo approach, which uses a likelihood sampling procedure based on weighted importance layers of conservation interest (similar to Marxan's) to evaluate alternative protected area sizing and placement. The two approaches are alternative options in a common spatial optimization module, which uses the time- and spatial dynamic Ecospace model for the evaluations. The optimizations are implemented as components of the Ecopath with Ecosim approach and software. In a case study, we find that there can be protected area zoning that will increase economic and social factors, without causing ecological deterioration. We also find a tradeoff between including cells of special conservation interest and the economic and social interest, and while this does not need to be a general feature, it points to the use of modeling techniques to evaluate the tradeoffs.

The most widely used approach for spatial planning with a conservation perspective is the Marxan approach and software, (http://www.uq.edu.au/marxan/) developed primarily by Hugh Possingham and colleagues at the University of Queensland. Marxan is a very flexible approach capable of incorporating large data sources and use categories, it is computationally efficient, and lends itself well to enabling stakeholder involvement in the site selection process.

We view the new importance layer sampling procedure as complimentary to the Marxan approach in that its strong side, through the underlying trophic modeling background is in evaluating ecological processes, including spatial connectivity and predicting future states – topics that were not well covered in the original Marxan analysis. In doing so, we, however, involve a rather complicated dynamic model, even if user-friendly, and this unavoidably has a cost. We therefore advocate that the two approaches, with their given advantages and limitations, be applied in conjunction – using two sources to throw light at a problem from different angles, beats one, any time. We have in order to facilitate such comparative studies developed a two-way bridge between Marxan and EwE, enabling exchange of spatial information and of optimization results between the two approaches.

Objective function

Table 1. Objective function employed for spatial optimization. Each objective is given a weighting factor, and the optimization seeks to optimize the summed, weighted objectives.

Objective	Description
Profit	Estimated by fleet, and summed over all such
Jobs	Estimated from value of fisheries, and relative number of jobs/value
Mandated rebuilding	A minimum acceptable level, by group
Ecosystem structure	Default values are based on biomass/productivity ratios expressing average longevity, weighted by group
Biomass diversity	Biomass evenness among groups
Boundary weight	Estimated as total boundary length over the protected area size. Relates to spatial connectivity

We employ an objective function for the optimizations that corresponds to the objective function used in the policy optimization module of EwE. This module uses a non-linear search routine to find a combination of effort by fishing fleets that will maximize the objective function.

The objective function in turn includes ecological, economic and social indicators, even legal constraints if pertinent, through considering profit, number of jobs, stock rebuilding, and two ecological measures. For the spatial optimizations we add a further indicator in form of a boundary weight factor (see Table 1).

The profit objective is calculated by summing revenue across all fleets, and subtracting the cost for operating. Cost is considered a linear function of effort with a fixed cost added. The following calculation,

$$R_t = \sum_f \sum_i (F_{fi} \cdot B_i \cdot V_{fi}) - \sum_f (E_f \cdot C_{v,f} - C_{p,f}) \tag{1}$$

is performed for each time t step to estimate the revenue R_t, with F_{fi} being the fishing mortality for group i caused by fleet f, B_i is the biomass of i, and V_{fi} is the ex-vessel value per unit weight of i caught by f. E_f is the relative effort for f, the $C_{v,f}$ is variable cost per unit effort for f, and $C_{p,f}$ is the fixed cost for fleet f.

The calculations in Eq. 1 are, as indicated, performed for each time step, with benefit summed over time. We, however, discount future values based on either a traditional discount rate, or an inter- generational discount rate[2], based on user preference.

As a social indicator, we use the number of jobs over time J_t created in the ecosystem, and we estimate this for each time step t from the landed value of the exploited group times the relative number of jobs per unit value N_i, or $J_t = \sum_f F_{fi} B_i V_{fi} N_i$. Similar to the profit objective, we discount the number of jobs over time.

We estimate the mandated rebuilding objective M_t for each time step t from

$$M_t = \sum_i B_i / B_{I*} \tag{2}$$

where B_i* is the baseline Ecopath biomass for group i, and equals the group biomass B_i if B_i is lower than the mandated biomass, $B_{m,i}$ for the group, and $B_{m,i}$ if it is not. The mandated rebuilding objective can be used to set "Minimum Biological Acceptable Levels" (or MBAL as commonly used) by giving this objective a high weighting. As long as the biomass is above MBAL the objective won't matter, but should it get below MBAL, it will! By setting high mandated biomasses $B_{m,i}$ for a group it can also be used to capture "existence values," e.g., of marine mammals of interest for a whale watching industry. We do not discount the mandated rebuilding structure over time.

The ecosystem structure objective is meant to capture that mature (K-type) ecosystems tend to be dominated by long-lived species and individuals[3]. We seek to capture this characteristic through the inverse production/biomass ratio, estimating for each time step

$$S_t = \sum_i B_i \cdot S_i \tag{3}$$

where S_t is the overall ecosystem structure measure, and S_i the ecosystem structure factor for i. We provide default values for S_i in form of the inverse P/B_i ratios (unit, year), supplied as part of the basic parameterization of the Ecopath model. To avoid unduly influence by very short-lived species we have (arbitrarily) set S_i to 0 for groups with an average lifespan of less than a year, (i.e. groups whose P/B_i is less than 1 year^{-1}).

The ecosystem structure objective is not discounted over time; having long-lived species in the future being deemed as important as having them now.

As a measure of biomass diversity, we use either the Shannon index or a modified version of Kempton's Q_{75} index, both of which originally were developed to describe species diversity[4]. We here used a biomass diversity indicator following Ainsworth and Pitcher[5] (2006), albeit slightly modified. We estimate the biomass diversity index Q_{75} from

$$Q_{75} = S/(2\log(N_{0.25-S}/N_{0.75-S})) \tag{4}$$

here S is the number of functional groups, and N_{i-S} is the biomass of the iS^{th} most common group, using a weighted average of the two closest group if iS is not an integer. The biomass diversity index describes the slope of a cumulative group abundance curve. As a sample with high diversity (evenness) will have a low slope, we reverse the index and express it relative to index value from the Ecopath base run $Q*_{75}$, that is

$$Q'_{75} = (2 - Q_{75})/Q^*_{75} \tag{5}$$

.

We truncate the index in the extreme and unlikely case that Q_{75} would more than double from the base run. We only include higher trophic level groups ($TL>3$) in the calculation of the biomass diversity index – should this, for models with only few functional groups, lead to less than 10 groups being included in the calculations. We, however, base the calculations on all living groups. As for the other ecological indicators, we do not discount future index values.

The final element in the objective function represents spatial connectivity, expressed through the boundary weight factor, L is estimated as $L = \sum_a A_c / \sum_b l_b$ where the total protected area size A_c is summed over spatial cells c, and the boundary length is estimated by summing over all protected cell b the side lengths l_b that do not border another protected cell or land. The boundary weight factor is similar to the Boundary Length Modifier that is used in Marxan to impact spatial fragmentation.

With the elements of the objective function being defined, we can now obtain the overall objective function measure O from

$$O = w_R \cdot R + w_J \cdot J + w_M \cdot M + w_S \cdot S + w_Q \cdot Q'_{75} \tag{6}$$

Where each of the objective weighting factors, w, can assume any value, including zero, which is used for measures that are ignored in a given optimization. We use the objective function measure for both of the optimization methods described below.

Seed cell selection procedure

This optimization method is based on a previous study[67], in which a very simple optimization scheme was used to evaluate tradeoff between proportion of area protected and the ecosystem-level objective function. We have modified the previous approach by securing a better program flow, and notably by changing the objective function from considering only profit from fishing and existence value of biomass groups to the more detailed function described above (Eq. 6).

The procedure takes as its starting point the designation of one, more, or all spatial cells as "seed cells", i.e. cells that are to be considered as potential protected cells in the next program iteration. The procedure will then run the Ecospace model repeatedly between two time steps, closing one of the seeds cells in each run, while storing the ecosystem objective function value. The seed cell that results in the highest objective function is then closed for fishing, and its four neighboring cells (above, below, and to either side) are then turned into seed cells, unless they are so already, or already are protected, or are land cells. This procedure will continue until all cells are protected.

The time over which the selection procedure is run is chosen dependent on the application. Typically, an ecosystem model is initially developed and tuned using time series data to cover a certain time period, e.g., from 1980 to 2020. Subsequently, the model is used in a scenario development mode to evaluate for instance protected area placement covering the period 2020-2040.

The major result from the seed cell selection procedure is an evaluation of the trade-off between size of protected area, and each of the objectives in Eq. 6. This can, for instance, be used to consider what proportion of the total area to close in subsequent, more detailed analysis based on importance layer sampling.

Importance layer sampling procedure

An advantage of the seed cell modeling approach described above is that it allows a comprehensive overview of the tradeoff between proportion of area closed to fishing, and the ecological, social, and economic benefit and costs of the closures. This is done, based on the information already included in the EwE modeling approach, with no new information being needed. While this may be an advantage from one perspective, it does not allow use of other forms of information, notably in form of geospatial data, such as, for instance, critical fish habitat layers from GIS.

To address this shortcoming, we have developed an alternative optimization routine for the Ecospace model, which uses spatial layers of conservation interest ("importance layers") to set likelihoods for spatial cells being considered for protection. The optimizations are performed using a Monte Carlo (MC) approach where the importance layers are used for the initial cell selection in each MC realization. The Ecospace model is then run, the objective function (Eq. 6) is evaluated, and the results, including which cells were protected, are stored for each run (see Figure 1).

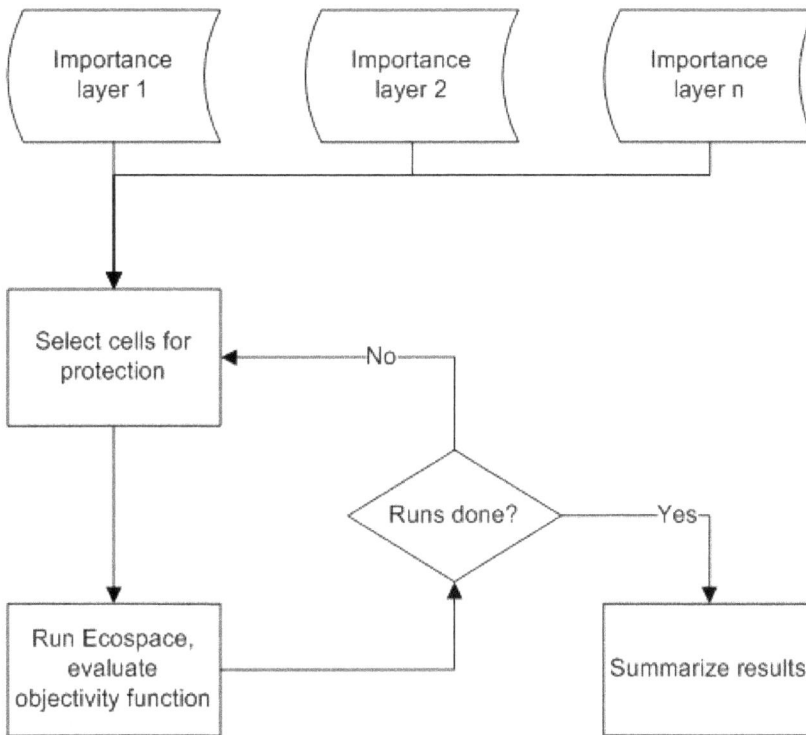

Figure 1. Logic of the importance layer sampling procedure. For each run a given percentage of all cells are protected based on weighted likelihood in importance layers. The evaluation of each run is done independently based on a defined objective function.

The importance layers are defined as raster layers, with dimensions similar to the base map layers in the underlying Ecospace model, i.e. they are rectangular cells in a grid with a certain number of rows and columns. Each cell in a given layer has a certain "importance" for conservation, expressed, e.g., as the probability of occurrence for an endangered species. For each importance layer λ, we initially scale the importance layer values to sum to unity, and then calculate an overall cell weighting w_c for each cell c from

$$w_c = \sum_\lambda w_\lambda \cdot C_{c,\lambda} \tag{7}$$

where w_{λ} are the importance layer weightings, and $C_{c,\lambda}$ the cell-specific, scaled importance layer values. In order to evaluate how well the importance layers are represented in each optimization run, we estimate

$$w'_\lambda = \frac{\sum_c w_\lambda C_{c',\lambda}}{\sum_c w_\lambda C_{c*,\lambda}} \tag{8}$$

where c' indicates cells selected in a given run, and c* is the cell with the highest weightings for the given layer. The layer-specific indicator w_λ' can obtain values in the range between 0 and 1.

For each optimization search, one has to select the proportion of water cells to protect in the runs, as well as how many times to repeat the Monte Carlo runs. It is possible to set the search routine up to iterate over a range of protection levels, e.g., from 10% to 100% protected in steps of 10%.

Similar to the seed cell selection procedure, we typically develop and tune the model to an initial time period, and then use the sampling procedure to evaluate scenarios for protected areas for a subsequent time period.

Ecospace can read raster files with spatial information such as importance layers or other Ecospace base map layers from comma separated text files (.csv), ESRI ASCII files (.asc), and ESRI shape files (.shp). The files need to have layers or columns with row and column numbers matching the Ecospace model. This capability is designed to allow straightforward exchange between the Ecospace modeling and Marxan analysis, with the constraint that it needs to be possible to represent the layers in raster form. The reading of the spatial files is described in more detail in the Spatial optimization chapter of the EwE User Guide.

Attribution This chapter is an edited extract from Christensen, V., Z. Ferdaña, J. Steenbeek. 2009. Spatial optimization of protected area placement incorporating ecological, social and economic criteria. Ecological Modelling 220:2583-2593 10.1016/j.ecolmodel.2009.06.029. Adapted with License Numbers 5757350148022 and 5757730967909 from Elsevier. The publication includes simple case studies for illustration.

Notes

1. An early version of the 'Ecoseed' approach was developed for Ecospace as part of a graduate student research project, (1) Beattie, A., 2001. A new model for evaluating the optimal size, placement, and configuration of marine protected areas. M.Sc, The University of British Columbia, Vancouver. (2) Beattie, A., Sumaila, U.R., Christensen, V. and Pauly, D., 2002. A model for the bioeconomic evaluation of marine protected area size and placement in the North Sea. Natural Resource Modeling, 15:413-437.

2. Sumaila, Ussif R. & Walters, Carl, 2005. Intergenerational discounting: a new intuitive

approach. Ecological Economics, 52(2): 135-142,

3. Odum, E.P. 1969. The strategy of ecosystem development. Science, 164:262-270. https://doi.org/10.1126/science.164.3877.262

4. Kempton, R.A., 2002. Species diversity. In: El-Shaarawi, A.H., Piegorsch, W.W. (Eds.), Encyclopedia of Environmetrics. John Wiley and Sons, Chichester, pp. 2086–2092.

5. Ainsworth, C.H., Pitcher, T.J., 2006. Modifying Kempton's species diversity index for use with ecosystem simulation models. Ecological Indicators 6, 623–630. http://dx.doi.org/10.1016/j.ecolind.2005.08.024

6. Beattie, A., 2001. A new model for evaluating the optimal size, placement, and configuration of marine protected areas. M.Sc. University of British Columbia, Vancouver.

7. Beattie, A., Sumaila, U.R., Christensen, V., Pauly, D., 2002. A model for the bioeconomic evaluation of marine protected area size and placement in the North Sea. Natural Resource Modeling 15, 413–437. https://doi.org/10.1111/j.1939-7445.2002.tb00096.x

53.

Coastal restoration

Kim de Mutsert

Coastal restoration projects, especially wetland recreation, benefit marine and coastal species by restoring habitat that serves as important nursery grounds[1][2]. The construction phase of the project and/or the environmental changes of the area under restoration are likely to have effects on the species currently residing in that area, and the impact of those changes need to be assessed. Coastal restoration projects are different from other construction projects that may affect species assemblages in that the long-term effects on the natural environment are aimed to be positive, and the future without restoration is likely to negatively affect coastal species over the long term.[3]

Ecospace is uniquely equipped to assess the effects of these environmental changes on fish and shellfish communities, as it can evaluate relatively short-term effects (months to years) of the construction/environmental disruption, as well as the long-term effects (decades) of having a restored environment on coastal and marine species that make use of that environment (for part of their life) over generations. By comparing this outcome to a future without action, the difference between taking this action or not can be evaluated over the short-term and the long-term.

A notable example of an area where large restoration projects are occurring and are planned is the Mississippi River Delta in the United States. In addition to various other coastal restoration and protection projects, the construction of large sediment diversions is planned, which are floodgates at select locations along the lower Mississippi River designed to let river water and sediments back into wetlands that were cut off from freshwater inflow by river levees in recent history (CPRA[4]). The introduction of freshwater and sediment through these floodgates will alter the environment of the receiving estuaries by reducing salinity, and increasing turbidity, nutrient concentrations, and wetland acreage amongst other changes.[5][6] Ecospace models developed to evaluate potential effects on fish and fisheries of these projects were included in resource managers' decisions on diversion flow regime and location[7](Figure 1).

The models provided anticipated redistribution of species[8][9], and demonstrated the potential impact of sea level rise on the anticipated outcome. This approach needs a coupled modeling framework, since Ecospace will simulate the effects on fish and fisheries of the environmental change that occurs as a result of diversion openings and restoration projects, while the environmental change itself (e.g., salinity, amount of habitat) needs to be simulated by different models[10][11][12].

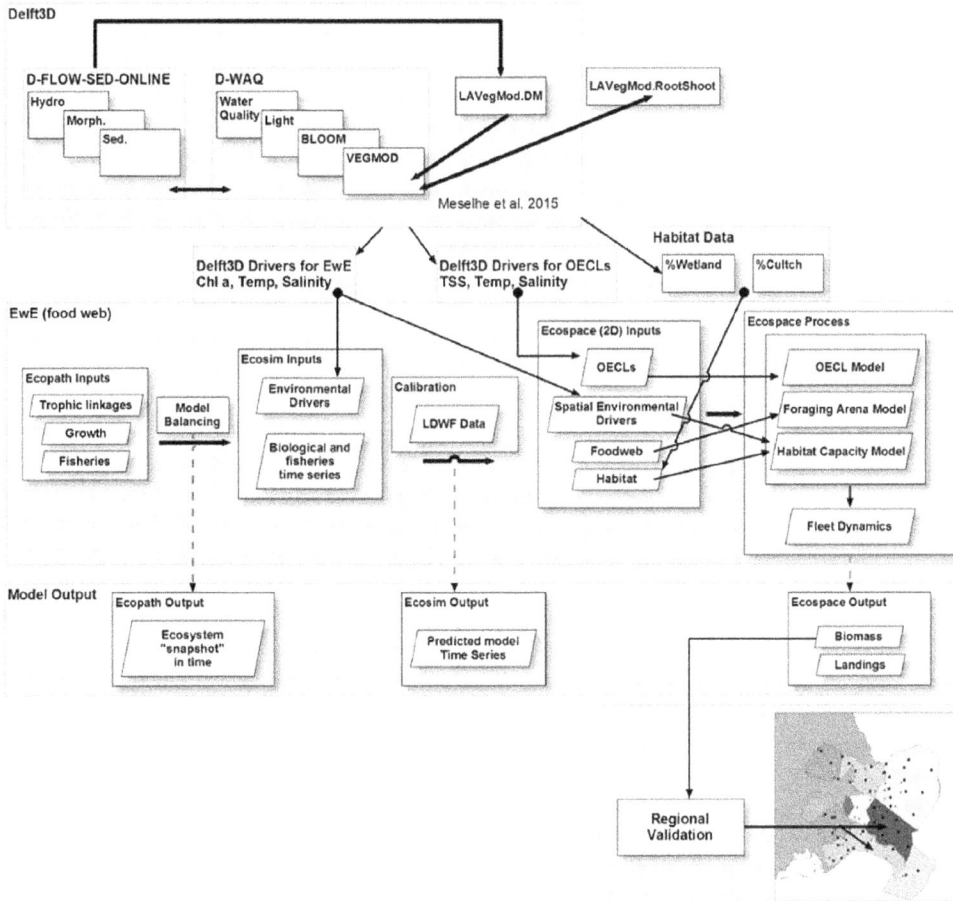

Figure 1 – Conceptual diagram of coupled modeling framework used in the Mississippi River Delta model[13].

Attribution

The chapter is based on de Mutsert K, Marta Coll, Jeroen Steenbeek, Cameron Ainsworth, Joe Buszowski, David Chagaris, Villy Christensen, Sheila J.J. Heymans, Kristy A. Lewis, Simone Libralato, Greig Oldford, Chiara Piroddi, Giovanni Romagnoni, Natalia Serpetti, Michael Spence, Carl Walters. 2023. Advances in spatial-temporal coastal and marine ecosystem modeling using Ecopath with Ecosim and Ecospace. Treatise on Estuarine and Coastal Science, 2nd Edition. Elsevier. https://doi.org/10.1016/B978-0-323-90798-9.00035-4, adapted with permission, License Number 5651431253138.

Rather than citing this chapter, please cite the source.

Notes

1. Minello, T.J., Able, K.W., Weinstein, M.P., Hays, C.G., 2003. Salt marshes as nurseries for nekton: testing hypotheses on density, growth and survival through meta-analysis. Marine Ecology Progress Series 246, 39–59. https://doi.org/10.3354/meps246039

2. Schulz, K., Stevens, P.W., Hill, J.E., Trotter, A.A., Ritch, J.L., Tuckett, Q.M., Patterson, J.T., 2020. restoration evaluated using dominant habitat characteristics and associated fish communities. PLOS ONE 15, e0240623. https://doi.org/10.1371/journal.pone.0240623

3. Rozas, L.P., Caldwell, P., Minello, T.J., 2005. The Fishery Value of Salt Marsh Restoration Projects. Journal of Coastal Research 37–50. https://www.jstor.org/stable/25736614

4. CPRA, 2017. Louisiana's Comprehensive Master Plan for a Sustainable Coast. Coastal Protection and Restoration Authority.

5. Baustian, M.M., Meselhe, E., Jung, H., Sadid, K., Duke-Sylvester, S.M., Visser, J.M., Allison, M.A., Moss, L.C., Ramatchandirane, C., Sebastiaan van Maren, D., Jeuken, M., Bargu, S., 2018. Development of an Integrated Biophysical Model to represent morphological and ecological processes in a changing deltaic and coastal ecosystem. Environmental Modelling & Software 109, 402–419. https://doi.org/10.1016/j.envsoft.2018.05.019

6. Das, A., Justic, D., Inoue, M., Hoda, A., Huang, H., Park, D., 2012. Impacts of Mississippi River diversions on salinity gradients in a deltaic Louisiana estuary: Ecological and management implications. Estuarine, Coastal and Shelf Science 111, 17–26. https://doi.org/10.1016/j.ecss.2012.06.005

7. De Mutsert, K., Lewis, K., Milroy, S., Buszowski, J., Steenbeek, J., 2017. Using ecosystem modeling to evaluate trade-offs in coastal management: Effects of large-scale river diversions on fish and fisheries. Ecological Modelling 360, 14–26. https://doi.org/10.1016/j.ecolmodel.2017.06.029

8. De Mutsert et al. 2017, *op. cit.*

9. De Mutsert, K., Lewis, K.A., White, E.D., Buszowski, J., 2021. End-to-End Modeling Reveals Species-Specific Effects of Large-Scale Coastal Restoration on Living Resources Facing Climate Change. Front. Mar. Sci. 8. https://doi.org/10.3389/fmars.2021.624532

10. Baustian, M.M., Meselhe, E., Jung, H., Sadid, K., Duke-Sylvester, S.M., Visser, J.M., Allison, M.A., Moss, L.C., Ramatchandirane, C., Sebastiaan van Maren, D., Jeuken, M., Bargu, S., 2018. Development of an Integrated Biophysical Model to represent morphological and ecological processes in a changing deltaic and coastal ecosystem. Environmental Modelling & Software 109, 402–419. https://doi.org/10.1016/j.envsoft.2018.05.019

11. Meselhe, E., Wang, Y., White, E., Jung, H., Baustian, M.M., Hemmerling, S., Barra, M., Bienn, H., 2020. Knowledge-Based Predictive Tools to Assess Effectiveness of Natural and Nature-Based Solutions for Coastal Restoration and Protection Planning. Journal of Hydraulic Engineering 146, 05019007. https://doi.org/10.1061/(ASCE)HY.1943-7900.0001659

12. White, E.D., Meselhe, E., Reed, D., Renfro, A., Snider, N.P., Wang, Y., 2019. Mitigating the Effects of Sea-Level Rise on Estuaries of the Mississippi Delta Plain Using River Diversions. Water 11, 2028. https://doi.org/10.3390/w11102028

13. Reproduced under CC BY-NC-ND 4.0 DEED from De Mutsert et al. 2017, *op. cit*

IX

Tracking Persistent Pollutants

54.

Introduction and dynamics

Shawn Booth; Jeroen Steenbeek; and Sabine Charmasson

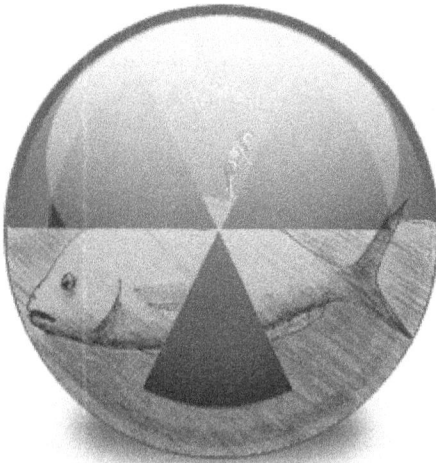

Ecotracer is a useful tool within the Ecopath with Ecosim (EwE) modelling approach to track radioisotopes, contaminants, persistent pollutants, or stable isotopes through a food web model. After achieving a mass-balanced Ecopath model, Ecotracer can be used with the Ecosim (time dynamic) or Ecospace (spatial-temporal dynamic) to track the flow of the pollutant through the modelled ecosystem.

EwE uses a mass-balance approach, and the flows of a contaminant due to predator/prey interactions are tracked within the underlying Ecopath model. However, Ecotracer also needs parameters for groups based on a kinetic toxicology approach to estimate initial conditions. However, similar to Ecopath, Ecotracer can become dynamic through either the use of Ecosim or Ecospace to follow the changes of a contaminant that has different temporal inputs or to variations in temporal spatial concentrations in the water column.

The purpose of this manual is to give a full description of the Ecotracer approach including the dynamic equations that describe the basis for the input parameters, and to familiarize users with the various interfaces for the inputs and outputs. Simulation scenarios are also given to allow users to become more familiar with Ecotracer, and instructions are given on how to navigate through the different interfaces used.

Introduction

Ecotracer is a sub-routine in the Ecopath with Ecosim (EwE) modeling framework[1] [2] that allows the modeller to follow a contaminant or stable isotope in modelled functional groups and the environment in a balanced Ecopath model.

Many EwE models have been made that focus on fishery-related questions, but here the focus is on how the Ecotracer routine is used within EwE to trace contaminants such as radionuclides through an aquatic ecosystem. EwE consists of three routines: Ecopath which is a mass balance interpretation of an ecosystem where, in essence, the production of a group in the model is equal to its consumption; Ecosim allows the user to build in time dynamics to the Ecopath model for events such as changes in contaminant loading to an ecosystem; and Ecospace which allows for the spatial-temporal resolution for such events as the effects of change in loading on marine organisms that result from organisms inhabiting different spatial areas or habitat types that have different environmental concentrations through time.

Typical applications of Ecotracer have been for contaminants such as mercury[3], ^{14}C[4][5], ^{137}Cs[6][7], and PCBs[8] that can have detrimental impacts on human and environmental health. The use of Ecotracer can help to estimate the amount of contaminant or concentration in a group/species of interest, spatial differences in concentration within the same functional groups if Ecospace is used, the fluxes between groups due to trophic interactions, and the importance of diet versus direct environmental uptake. It can also help to estimate functional groups' concentrations when such data are lacking (i.e., have a starting value of zero) and make forward projections based on changing environmental concentrations. Concentration levels are an important aspect for environmental and human health as, in conjunction with consumption rates, they determine exposure levels that may have detrimental effects. Regulatory limits on the concentration in aquatic products destined for human consumption may also affect trade and fisheries opportunities.

Ecotracer requires a balanced Ecopath model to follow the contaminant, radioisotopes or stable isotopes in the model groups, and environment (e.g., water concentration). Ecotracer when used with Ecosim can provide estimates to important ecotoxicological questions such as,

1. what could be the expected group concentrations if the environmental concentration did not change?

2. what could be the expected results in group concentrations if the environmental concentration changed through time? and

3. is there an effect on concentration levels as a result of changing underlying Ecopath parameters such as fishing mortality?

The first question is useful if many functional groups in the model lack concentration data. The second question can be important to estimate resulting concentrations in biota if the input into the environment changes. The third can help to understand contaminant flows as a result of changes in the dynamics of the underlying structure of an ecosystem.

Ecotracer when used with Ecospace can help to answer whether there are differences in the same species that occur over a large geographic area, and whether different environmental concentrations in different areas impact the resulting concentrations in organisms. In this case, a two-dimensional representation of the model area is made which has a user defined spatial resolution (i.e., grid cells). Spatial environmental concentrations can be driven by effluents being released as a point source, or from atmospheric deposits that change over space and time. Effluents released from a point source would be affected by currents resulting in different spatial and temporal distributions, and atmospheric releases could be affected by different levels of releases due to industrial activity through time or accident scenarios as well as currents. In the case of large-scale accidents, such as the Dai-ichi nuclear accident at Fukushima, spatial differences can result from both point sources and differing atmospheric deposits both of which occurred through time.

Ecotracer dynamics

Ecotracer simulates the contaminant fluxes and resulting amounts and concentrations using a modified transfer contaminant model (e.g.,[9][10]), and applies it to both the environment and biota. Resulting changes at any time step are dependent upon the gains and losses in functional groups and are described in Walters and Christensen[11] as,

$$\frac{dA_i(t)}{dt} = \alpha_i - \beta_i A_i(t) \tag{1}$$

where α_i represents the gains (Bq·year^{-1}) in each functional group i, β_i represents the rate losses (year^{-1}) to each functional group, and A_i represents the amount (e.g., Bq) in each functional group i. This general formulation allows different measurement units of substances (e.g., Bq or µg) to be tracked in the modelled environment, and the resulting concentrations (e.g., Bq·t^{-1}) are computed separately using the biomass output in Ecopath and Ecosim.

The environmental compartment concentrations are also calculated by tracking the gains and losses in the cells representing the environment,

$$\frac{dC_o(t)}{dt} = \alpha_o - \beta_o C_o(t) \tag{2}$$

where C_o represents the environmental concentration (e.g., Bq·km^{-2}), α_o, represents the gains and losses in each environmental cell o, and β_o represents the rate losses (year^{-1}) in each cell.

Environment

Gains in the environment originate from the release of contaminants into the environment as a base inflow rate, and from the excretion from organisms. Losses originate from the direct uptake from the environment by organisms, physical decay rates, and base volume exchange. In Ecospace, the environment can be represented by multiple grid cells and thus the gains and losses can be considered to be for each environmental compartment o, such that,

$$\alpha_o = BI_o + \sum_{i=1}^{n} m_i A_i \tag{3}$$

where BI_0 is the base inflow rate (Bq·km^{-2}·year^{-1}) to a grid cell, and m_iA_i are the excretory products for each functional group within each grid cell.

Losses from the environment are due to biological, physical decay processes, environmental volume changes, and direct uptake by organisms, such that,

$$\beta_o C_o = (d_i + V_i)C_o + \sum_{i=1}^{n} u_i B_i C_o \tag{4}$$

where d_i represents the physical decay rate (year^{-1}), V_i represents the base volume exchange loss (year^{-1}), and the second term ($u_i B_i C_0$) represents the total uptake rate by all functional groups (see below). Temporal changes to the environmental concentration (C_0) can be made by applying a forcing function to the base inflow rate, through a contaminant concentration driver file (Table 1), or by current/advection fields.

Biota

In biota, intake amounts (e.g., Bq·year-1) result from direct uptake rates (i.e., respiration) the fraction retained from trophic interactions (i.e., diet), and immigration. i.e.,

$$\alpha_i = u_i B_i C_o + AE_i \sum_{i=1}^{} Q_{ij} \frac{A_j}{B_j} + c_i I_i \tag{5}$$

where, C_0 represents the environmental concentration (Bq·km^{-2}), B_i is the biomass (t) of group i, u_i represent the intake/biomass/environmental concentration/year (km^2·t^{-1}·year^{-1}); AE_i is the assimilation efficiency for each group, Q_{ji} is the consumption rate (t·year^{-1}) of group j by group i, A_j is the amount of substance in a group (e.g., Bq), B_j is the prey biomass of each prey item j (Bq·t^{-1}); c_i is the group biomass concentration (Bq·t^{-1}) and I_i is the immigrating biomass (t·year^{-1}).

The losses from a group ($\beta_i C_i$) are attributed to predation, fisheries, other mortality, excretion and decay, i.e.,

$$\beta_i C_i = (\sum_{j=pred} (\frac{Q_{ij}}{B_i}) + F_i + MO_i + E_i + m_i + d_i)C_i \tag{6}$$

where Q_{ij} is the rate of consumption (t·year^{-1}) of group i due to predation by j, F_i is the fishing mortality rate (year-1), MO_i (year^{-1}) is other mortality rate (i.e., non-predation mortality), E_i is the emigrating biomass rate (year^{-1}), m_i (year^{-1}) is the excretion and/or metabolic rate, and d_i (year^{-1}) is the physical decay rate. These rates are multiplied by C_i the amount of contaminant (Bq) in each group i. Excretory products that are released from tissues to the environment are added to the environmental concentration.

The solution for finding the equilibrium amount of contaminant in a primary producer with the resulting concentration only being due to direct uptake, losses to due predation, other mortality, metabolism, and decay is given as,

$$C_{i,eq} = \frac{u_i B_i C_o}{\sum\limits_{j=pred} (\frac{Q_{ij}}{B_i}) + MO_i + m_i + d_i} \tag{7}$$

whereas for other groups an additional term must be accounted for due to the group's prey items; in these cases the equilibrium solution can be defined as,

$$C_{i,eq} = \frac{u_i B_i C_o + AE_i \frac{Q_{ji}}{B_i}}{\sum\limits_{j=pred} (\frac{Q_{ij}}{B_i}) + MO_i + m_i + d_i} \tag{8}$$

The Ecotracer approach is dynamic and extends the basic concentration ratio (CR) approach, but the CR approach is contained within it as,

$$CR_i = \frac{(A_i/B_i)}{C_o} = \frac{u_i + AE_i \frac{Q_{ji}}{B_i} CR_j}{Z_i + m_i} \tag{9}$$

The amount of contaminant in the detritus compartment originates from the unassimilated consumption resulting from predation, as well as non-predation mortality. Thus, groups feeding on detritus will have exposure levels associated with the contributions from the fraction of unassimilated consumption from all groups. Initial concentrations in the biota and environment are also input parameters that can be used if data is available. For groups lacking contaminant data from field studies or literature data, the model is able to estimate concentration or burdens in the groups, leading to the ability to estimate risk through time and make comparisons to regulatory limits.

Attribution

This work was funded by the Institut de Radioprotection et de Sûreté Nucléaire (IRSN) and the French program Investissement d'Avenir run by the National Research Agency (AMORAD project, grant ANR-11-RSNR-0002, 2013-2022)

.

Notes

1. Christensen, V., Walters, C.J., 2004. Ecopath with Ecosim: methods, capabilities and limitations. Ecological Modelling, Placing Fisheries in their Ecosystem Context 172, 109–139. https://doi.org/10.1016/j.ecolmodel.2003.09.003

2. Walters, W.J., Christensen, V., 2018. Ecotracer: analyzing concentration of contaminants and radioisotopes in an aquatic spatial-dynamic food web model. Journal of Environmental Radioactivity 181, 118–127. https://doi.org/10.1016/j.jenvrad.2017.11.008

3. Booth, S., Zeller, D., 2005. Mercury, Food Webs, and Marine Mammals: Implications of Diet and Climate Change for Human Health. Environmental Health Perspectives 113, 521–526. https://doi.org/10.1289/ehp.7603

4. Sandberg, J., Kumblad, L., Kautsky, U., 2007. Can ECOPATH with ECOSIM enhance models of radionuclide flows in food webs? – an example for 14C in a coastal food web in the Baltic Sea. Journal of Environmental Radioactivity 92, 96–111. https://doi.org/10.1016/j.jenvrad.2006.09.010

5. Tierney, K.M., Heymans, J.J., Muir, G.K.P., Cook, G.T., Buszowski, J., Steenbeek, J., Walters, W.J., Christensen, V., MacKinnon, G., Howe, J.A., Xu, S., 2018. Modelling marine trophic transfer of radiocarbon (14C) from a nuclear facility. Environmental Modelling & Software 102, 138–154. https://doi.org/10.1016/j.envsoft.2018.01.013

6. Walters & Christensen. 2018, *op. cit.*

7. Booth, S., Walters, W.J., Steenbeek, J., Christensen, V., Charmasson, S., 2020. An Ecopath with Ecosim model for the Pacific coast of eastern Japan: Describing the marine environment and its fisheries prior to the Great East Japan earthquake. Ecological Modelling 428, 109087. https://doi.org/10.1016/j.ecolmodel.2020.109087

8. Booth, S., Cheung, W.W.L., Coombs-Wallace, A.P., Zeller, D., Christensen, V., Pauly, D., 2016. Pollutants in the seas around us, in: Pauly, D., Zeller, D. (Eds.), Global Atlas of Marine Fisheries: A Critical Appraisal of Catches and Ecosystem Impacts. pp. 152–170.

9. Landrum, P.F., Lydy, M.J., Lee, H., 1992. Toxicokinetics in aquatic systems: Model comparisons and use in hazard assessment. Environ Toxicol Chem 11, 1709–1725. https://doi.org/10.1002/etc.5620111205

10. Thomann, R.V., 1981. Equilibrium Model of Fate of Microcontaminants in Diverse Aquatic Food Chains. Can. J. Fish. Aquat. Sci. 38, 280–296. https://doi.org/10.1139/f81-040

11. Walters & Christensen, 2018, *op. cit.*

55.

Ecotracer applications

Ecotracer is the unofficial fourth module of the Ecopath with Ecosim software, designed for tracking persistent contaminants in food webs. Ecotracer requires a balanced Ecopath model to trace the contaminant in model groups/species of the model and in the environment (e.g., water concentration)[1][2].

After achieving a mass-balanced Ecopath model of a specific ecosystem, Ecotracer simulates the flows of a contaminant due to predator/prey interactions following Ecopath parameters, the temporal changes of these flows through Ecosim and spatial-temporal dynamic of the contaminant through Ecospace. Ecotracer requires contaminant specific parameters for the modelled functional groups based on a kinetic toxicology approach to estimate initial conditions.

Typical applications of Ecotracer have been for contaminants that can have potential detrimental impacts on human and environmental health including bio-accumulating heavy metal such as mercury[3], radioisotopes and stable isotopes[4][5][6], polychlorinated biphenyls (PCBs)[7] and more recently microplastic[8][9].

Ecotracer estimates the concentration of a contaminant in modeled groups and computes temporal (Ecosim) and spatial (Ecospace) build-ups in concentration. Ecotracer simulates the contaminant fluxes and resulting concentrations in each group using a modified transfer contaminant model[10][11] that applies to both the environment and biota. In practice, Ecotracer calculates the contaminant concentration as trade-off between "gains" and "losses" for the environment and all groups/species in the model (Table 1). The use of a food web modeling approach allows disentangling contaminant fluxes through groups considering their direct uptake from the environment as well as through trophic interactions. Ecotracer can also estimate functional groups' contaminant concentrations when data of the initial conditions are lacking, and make forward projections based on changing environmental concentrations.

Table 1 – Gains and losses that can be accounted for during contaminant tracing using Ecotracer

Environment (e.g., seawater)

Gains	Losses
Contaminant inflow rate to environment	Contaminant decay rates in the environment
The sum of all contaminant excretory products from all the living groups/species	Contaminant outflow rate
	The sum of all contaminant uptake rates by all groups/species

For each group/species

Gains	Losses
Direct contaminant uptake rates from the environment	Contaminant decay rates in the group
Indirect contaminant uptake rates from trophic interactions	Contaminant excretion rates

Attribution

The chapter is based on de Mutsert et al.[12], adapted with permission, License Number 5651431253138. Rather than citing this chapter, please cite the source.

Notes

1. Christensen, V., Walters, C.J., 2004. Ecopath with Ecosim: methods, capabilities and limitations. Ecological Modelling, Placing Fisheries in their Ecosystem Context 172, 109–139. https://doi.org/10.1016/j.ecolmodel.2003.09.003

2. Walters, W.J., Christensen, V., 2018. Ecotracer: analyzing concentration of contaminants and radioisotopes in an aquatic spatial-dynamic food web model. Journal of Environmental Radioactivity 181, 118–127. https://doi.org/10.1016/j.jenvrad.2017.11.008

3. Booth, S., Zeller, D., 2005. Mercury, Food Webs, and Marine Mammals: Implications of Diet and Climate Change for Human Health. Environmental Health Perspectives 113, 521–526. https://doi.org/10.1289/ehp.7603

4. Sandberg, J., Kumblad, L., Kautsky, U., 2007. Can ECOPATH with ECOSIM enhance models of radionuclide flows in food webs? – an example for 14C in a coastal food web in the Baltic Sea. Journal of Environmental Radioactivity 92, 96–111. https://doi.org/10.1016/j.jenvrad.2006.09.010

5. Tierney, K.M., Heymans, J.J., Muir, G.K.P., Cook, G.T., Buszowski, J., Steenbeek, J., Walters,

W.J., Christensen, V., MacKinnon, G., Howe, J.A., Xu, S., 2018. Modelling marine trophic transfer of radiocarbon (14C) from a nuclear facility. Environmental Modelling & Software 102, 138–154. https://doi.org/10.1016/j.envsoft.2018.01.013

6. Booth, S., Walters, W.J., Steenbeek, J., Christensen, V., Charmasson, S., 2020. An Ecopath with Ecosim model for the Pacific coast of eastern Japan: Describing the marine environment and its fisheries prior to the Great East Japan earthquake. Ecological Modelling 428, 109087. https://doi.org/10.1016/j.ecolmodel.2020.109087

7. Booth, S., Cheung, W.W.L., Coombs-Wallace, A.P., Zeller, D., Christensen, V., Pauly, D., 2016. Pollutants in the seas around us, in: Pauly, D., Zeller, D. (Eds.), Global Atlas of Marine Fisheries: A Critical Appraisal of Catches and Ecosystem Impacts. Island Press. pp. 152–170.

8. Boyer, J., Rubalcava, K., Booth, S., Townsend, H., 2022. Proof-of-concept model for exploring the impacts of microplastics accumulation in the Maryland coastal bays ecosystem. Ecological Modelling 464, 109849. https://doi.org/10.1016/j.ecolmodel.2021.109849

9. Ma, Y., You, X., 2021. Modelling the accumulation of microplastics through food webs with the example Baiyangdian Lake, China. Science of The Total Environment 762, 144110. https://doi.org/10.1016/j.scitotenv.2020.144110

10. Thomann, R.V., 1981. Equilibrium Model of Fate of Microcontaminants in Diverse Aquatic Food Chains. Can. J. Fish. Aquat. Sci. 38, 280–296. https://doi.org/10.1139/f81-040

11. Landrum, P.F., Lydy, M.J., Lee, H., 1992. Toxicokinetics in aquatic systems: Model comparisons and use in hazard assessment. Environ Toxicol Chem 11, 1709–1725. https://doi.org/10.1002/etc.5620111205

12. De Mutsert K, Marta Coll, Jeroen Steenbeek, Cameron Ainsworth, Joe Buszowski, David Chagaris, Villy Christensen, Sheila J.J. Heymans, Kristy A. Lewis, Simone Libralato, Greig Oldford, Chiara Piroddi, Giovanni Romagnoni, Natalia Serpetti, Michael Spence, Carl Walters. 2023. Advances in spatial-temporal coastal and marine ecosystem modeling using Ecopath with Ecosim and Ecospace. Treatise on Estuarine and Coastal Science, 2nd Edition. Elsevier. https://doi.org/10.1016/B978-0-323-90798-9.00035-4

56.

Modelling micro plastics

Chiara Piroddi; Natalia Serpetti; and William Walters

Plastic production in the EU has increased in the last 50 years. According to Penca,[1] 60-80% of total plastic-waste ends up in the oceans, suggesting that it will continue to grow if no waste management infrastructure improvements are put in place.[2] More than 90% of all plastic items (by number) found at sea belongs to microplastics (MP; items < 5 mm).[3]

Plastic pollution of the oceans is a very high priority topic in the context of different EU legislations such as the Urban Waste Water Treatment Directive (2008), the Marine Strategy Framework Directive (MSFD, 2008), the Waste Framework Directive (2011), the Plastic Strategy (2020), the Biodiversity Strategy for 2030 (2021) and the Zero Pollution Action Plan (2022). All these policies constitute important milestones of the roadmap initiated by the European Commission to achieve the European Green Deal (EC, 2020) which aspires to "protect the health and well-being of citizens from environment-related risks and impacts" and establish a toxic and plastic-free environment, deliver healthy and sustainable diets, and protect biodiversity.

For this, the European Commission (EC) Joint Research Centre (JRC) has developed an integrated modelling framework, called the Blue2 Modelling Framework (MF), to assess the impacts of diverse management strategies (including litter) on the status of EU freshwater and marine ecosystems. This framework incorporates models for freshwater quantity and quality, to recreate the conditions of EU rivers and lakes, as well as atmospheric forcing to capture atmospheric deposition of important chemical elements for marine ecosystems. At the core of the Blue2MF, there is an ocean model that consists of different modules. A hydrodynamic component, common for all European seas, a biogeochemical module and a high trophic level (HTL) module expertly customized for each EU marine region/ecosystem, and, a Lagrangian module used to simulate dispersion and accumulation patterns of floating litter (Figure 1). The Blue2MF can be integrated in different time-slices, from the 1970s to the present day, for hindcast simulations, and in forecasting mode (up to 2050), linked to the atmospheric conditions provided by IPCC-type global circulation models[4] The HTL module of the Blue2 MF is using the software EwE with all its components: Ecopath, Ecosim and Ecospace.

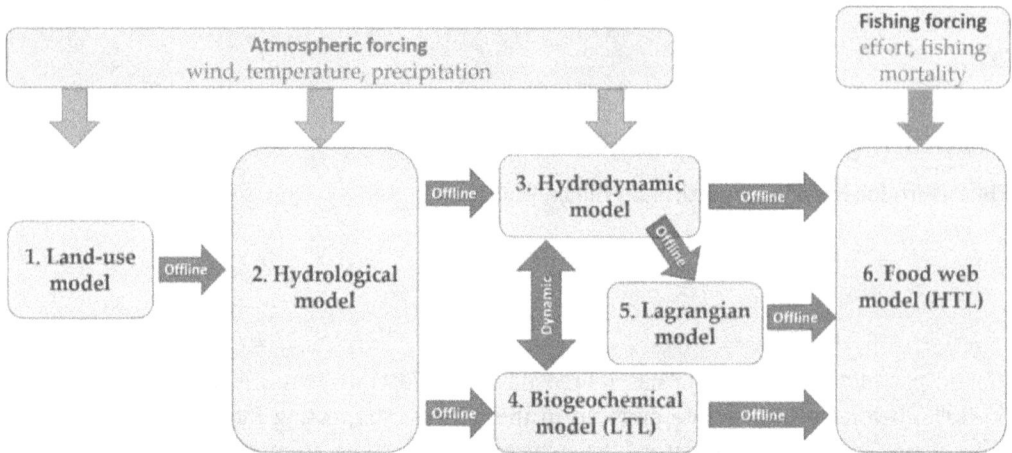

Figure 1. The Blue2 modelling framework used by EC-JRC for modelling environmental impacts and status.

The Blue2MF has also used the Ecotracer module of EwE for the Black and Mediterranean seas ecosystems to simulate and analyze the uptake of MP through the food web.

Among EU regional seas, these basins are particularly sensible to plastic pollution. In fact, their semi-enclosed nature, highly populated coasts,[5] large touristic and maritime activities, make them a concentration area from where floating litter could not escape.[6] MP ingestion by marine organisms is likely a major pathway for plastic in these ecosystems. Although MP are rapidly ingested and egested, the effects of MP ingestion in natural populations and their fate in marine food webs remain elusive. Without knowledge of retention and excretion rates of field populations, it is difficult to deduce ecological consequences[7] and assess the overall potential loss of energy when MP is consumed by the species of the food web.

Ecotracer calculates the amount of MP per unit biomass of each species in the ecosystem. These concentrations are of course depending on the MP concentration in the environment and varies depending on their diet (MP concentration in their preys), species direct absorption from the environment and species excretion rates. Within the Blue2MF Ecotracer module, the initial conditions of MP in the environment (concentration and basin inflow/outflow) as well as functional groups excretion rates were estimated from bibliography. A global database of species/MP ingestion was constructed for this purpose[8] and the models were then calibrated against observations of MP in the diet of all the functional groups.[9]

Results showed that, at steady state, in both ecosystems, primary consumers functional groups (benthic and pelagic) revealed the highest concentration of MP particles: they represented the species with the main MP pathways within the food web.[10] Future scenarios were run in Ecosim to simulate the impact of potential policies (10% and 50% reduction) aiming to reduce MP input in both basins, whilst Ecospace was used to identify hot-spots areas of co-occurrence between targeted sensitive species/functional groups, in terms of MP uptake, and floating particles, derived from the Blue2 Lagrangian module.[11]

Attribution: This chapter is in part adapted from Duteil et al. (2023).[12]

Notes

1. Penca, J. (2018). European Plastics Strategy: What promise for global marine litter? Marine Policy 97:197-201. https://doi.org/10.1016/j.marpol.2018.06.004

2. Jambeck, R.J., Geyer, R., Wilcox, C., Siegler, T.R., Perryman, M., Andrady, A., Narayan, R., Law, K.L. (2015). Marine pollution. Plastic waste inputs from land into the ocean. Science 347: 768-771. DOI: 10.1126/science.1260352

3. Eriksen, M., Lebreton, L.C.M., Carson, H.S., Thiel, M., Moore, C.J., Borerro, J.C., Galgani, F., Ryan, P.G., Reisser, J. (2014). Plastic Pollution in the World's Oceans: More than 5 Trillion Plastic Pieces Weighing over 250,000 Tons Afloat at Sea. PLOS ONE, 9(12): e111913. https://doi.org/10.1371/journal.pone.0111913

4. Stips, A. Dowell, M., Somma, F., Coughlan, C., Piroddi, C., Bouraoui, F., Macias, D., Garcia-Gorriz, E., Cardoso, A.C., Bidoglio, G. (2015). Towards an integrated water modelling toolbox. European Commission, Luxemburg.

5. Jambeck et al. (2015). *op. cit.* DOI: 10.1126/science.1260352

6. Ryan, P.G. (2013). Simple technique for counting marine debris at sea reveals steep litter gradients between the Straits of Malacca and the Bay of Bengal. Marine Pollution Bulletin, 69: 128-136. https://doi.org/10.1016/j.marpolbul.2013.01.016

7. Lusher, A. (2015). Microplastics in the Marine Environment: Distribution, Interactions and Effects. In Marine Anthropogenic Litter, pp. 245-307. Cham: Springer. https://doi.org/ 10.1007/978-3-319-16510-3_10

8. Serpetti N, Walters, W., Piroddi C., Garcia Gorriz E., Miladinova S., Macias D., Tracing microplastics up the EU marine food webs: implications for marine biodiversity and EU ecosystem services (PLASTIC-WEB) - Uptake of plastic by marine organism's database, Ispra: European Commission, 2022, JRC130033.

9. Serpetti, N., Walters, W., Piroddi, C., Garcia-Gorriz, E., Miladinova, S., Macias, D., Tracing microplastics up the EU marine food webs: implications for marine biodiversity and EU ecosystem services (PLASTIC-WEB) - Ecotracer modules setup for the Black and Mediterranean Seas, European Commission, Ispra, 2023 , JRC133312.

10. Serpetti, N., Walters, W., Piroddi, C., Garcia-Gorriz, E., Miladinova, S., Macias, D., Tracing microplastics up the EU marine food webs: implications for marine biodiversity and EU ecosystem services (PLASTIC-WEB) - Ecotracer modules setup for the Black and Mediterranean Seas, European Commission, Ispra, 2023, JRC133312

11. Serpetti, N., Walters, W., Piroddi, C., Garcia-Gorriz, E., Miladinova, S., Macias, D., Tracing microplastics up the EU marine food webs: implications for marine biodiversity and EU ecosystem services (PLASTIC-WEB) – Final reporting, European Commission, Ispra, 2023 (b), JRC134899.

12. Duteil, O., Macias Moy, D., Piroddi, C., Serpetti, N., Stips, A., Ferreira Cordeiro, N., Garcia Gorriz, E., Miladinova-Marinova, S., Parn, O., Polimene, L., Booth, S., Compa Ferrer, M., Dabrowski, T., Fuortibuonni, T., Gonzales-Fernandes, D., Laurent, C., Liubartseva, S., Suaria, G., Tekman, M., Tsiaras, K. and Walters, W., Report of the 5th meeting of the Network of Experts for ReDeveloping Models of the European Marine Environment, Publications Office of the European Union, Luxembourg, 2023, dx.doi.org/10.2760/114580, JRC133204.

Contributors

Authors

Villy Christensen

INSTITUTE FOR THE OCEANS AND FISHERIES, THE UNIVERSITY OF BRITISH COLUMBIA

https://oceans.ubc.ca/villy-christensen/

Carl Walters

INSTITUTE FOR THE OCEANS AND FISHERIES, THE UNIVERSITY OF BRITISH COLUMBIA

https://oceans.ubc.ca/carl-walters/

Contributors

Robert NM Ahrens

PACIFIC ISLANDS FISHERIES SCIENCE CENTER, NOAA

https://www.fisheries.noaa.gov/contact/robert-ahrens-phd

Jacob Bentley
NATURAL ENGLAND
https://www.researchgate.net/profile/Jacob-Bentley

Shawn Booth

Joe Buszowski
MOUNTAINSOFT
https://mountainsoft.net

David Chagaris
IFAS NATURE COAST BIOLOGICAL STATION. UNIVERSITY OF FLORIDA
https://sites.google.com/ufl.edu/dchagaris-fisheries-lab

Sabine Charmasson
INSTITUT DE RADIOPROTECTION ET DE
SÛRETÉ NUCLÉAIRE (IRSN)
https://www.researchgate.net/lab/LRTA-IRSN-Res-
Lab-on-Radionuclides-Transfer-in-Aquatic-ecosys-
tems-R-Gurriaran

Marta Coll
CSIC - INSTITUTO DE CIENCIAS DEL MAR (ICM)
https://martacollmarine.science

Sheila JJ Heymans
EUROPEAN MARINE BOARD
https://www.marineboard.eu/secretariat

KIm de Mutsert
SCHOOL OF OCEAN SCIENCE AND ENGINEER-
ING, THE UNIVERSITY OF SOUTHERN MISSIS-
SIPPI
https://demutsertlab.com/

Chiara Piroddi
EUROPEAN COMMISSION, JOINT RESEARCH
CENTRE
https://www.researchgate.net/profile/Chiara-Piroddi

Jeffrey Polovina
PACIFIC ISLANDS FISHERIES SCIENCE CENTER,
NOAA (RETIRED)
https://scholar.google.com/citations?user=n3DZWAkAAAAJ

Santiago de la Puente
NORWEGIAN INSTITUTE FOR WATER ANALYSIS
(NIVA)
https://www.niva.no/en/employees/santiago-de-la-puente-jeri

Natalia Serpetti
EUROPEAN COMMISSION, JOINT RESEARCH
CENTRE
https://www.researchgate.net/profile/Natalia-Serpetti
https://www.linkedin.com/in/dr-natalia-serpetti-89602312/?originalSubdomain=it

Jeroen Steenbeek
ECOPATH INTERNATIONAL INITIATIVE
https://ecopathinternational.org/team/

William Walters
PENNSYLVANIA STATE UNIVERSITY
https://www.researchgate.net/profile/William-Walters-6

Reviewer

Greig Oldford
FISHERIES AND OCEANS CANADA
https://goldford.github.io/

Illustrator

Jacob Bentley
NATURAL ENGLAND
https://www.researchgate.net/profile/Jacob-Bentley

www.ingramcontent.com/pod-product-compliance
Lightning Source LLC
Chambersburg PA
CBHW051752200326
41597CB00025B/4522